PROFILES IN MATHEMATICS

Ancient Mathematicians

Profiles in Mathematics:

Ancient Mathematicians

Rob Staeger

MORGAN
REYNOLDS
PUBLISHING
Greensboro, North Carolina

Profiles in Mathematics:

Alan Turing

Rene' Descartes

Carl Friedrich Gauss

Sophie Germain

Pierre de Fermat

Ancient Mathematicians

Women Mathematicians

PROFILES IN MATHEMATICS
ANCIENT MATHEMATICIANS

Morgan Reynolds Publishing, Inc., 620 South Elm Street, Suite 223
Greensboro, North Carolina 27406 USA

Library of Congress Cataloging-in-Publication Data

Staeger, Rob.
Ancient mathematicians / by Rob Staeger.
p. cm.
Includes bibliographical references and index.
ISBN-13: 978-1-59935-065-3
ISBN-10: 1-59935-065-3
1. Mathematicians--Biography. 2. Mathematicians--Greece--Biography.
3. Mathematics, Ancient. 4. Mathematics, Greek. I. Title.
QA28.S73 2007
510.92'238--dc22
[B]

2008007533

Printed in the United States of America

First Edition

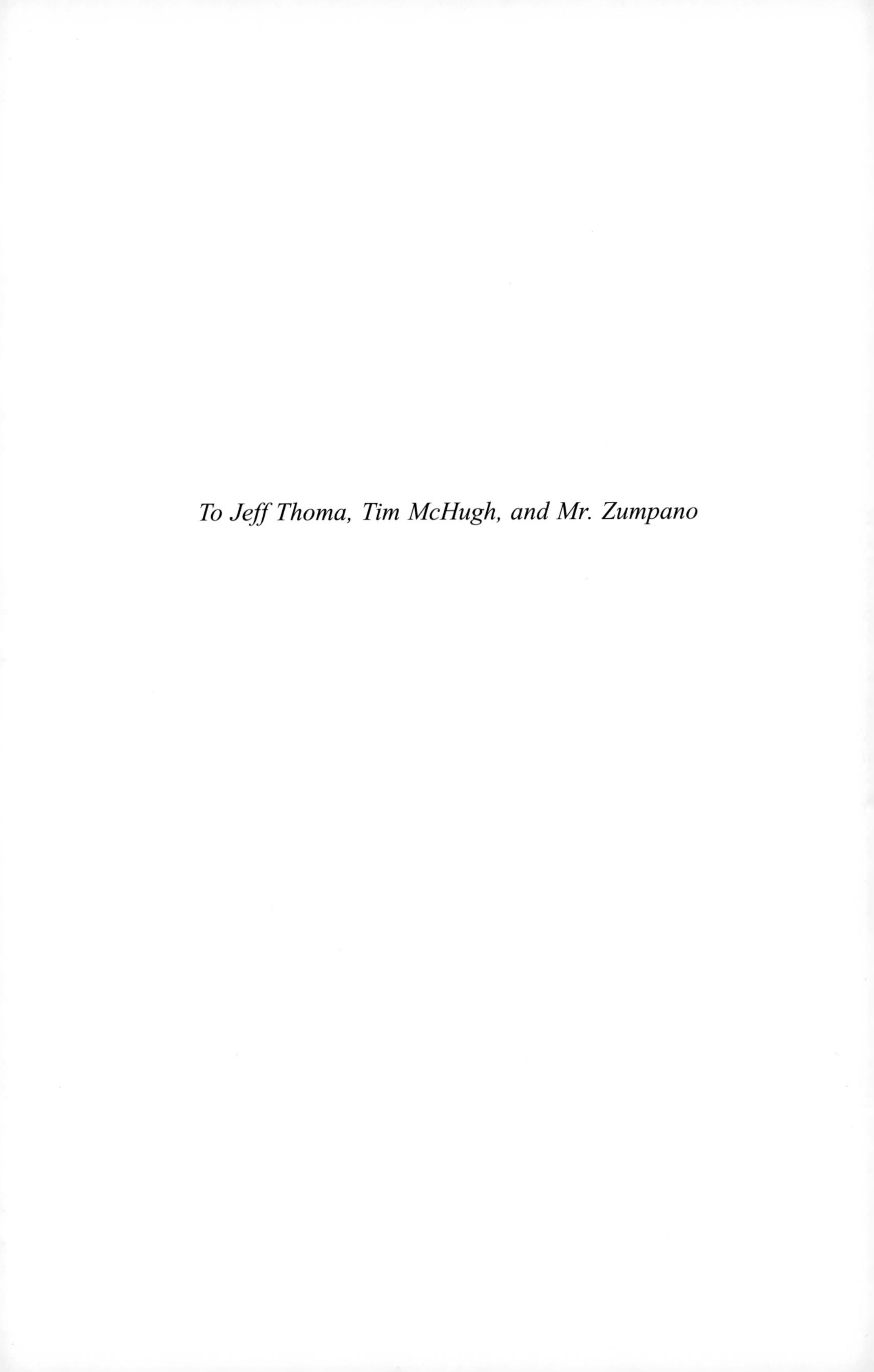

To Jeff Thoma, Tim McHugh, and Mr. Zumpano

Contents

Introduction

Mathematics gives us a powerful way to analyze and try to understand many of the things we observe around us, from the spread of epidemics and the orbit of planets, to grade point averages and the distance between cities. Mathematics also has been used to search for spiritual truth, as well as the more abstract question of what is knowledge itself.

Perhaps the most intriguing question about mathematics is where does it come from? Is it discovered, or is it invented? Does nature order the world by mathematical principles, and the mathematician's job is to uncover this underlying system? Or is mathematics created by mathematicians as developing cultures and technologies require it? This unanswerable question has intrigued mathematicians and scientists for thousands of years and is at the heart of this new series of biographies.

The development of mathematical knowledge has progressed, in fits and starts, for thousands of years. People from various areas and cultures have discovered new mathematical concepts and devised complex systems of algorithms and equations that have both practical and philosophical impact.

To learn more of the history of mathematics is to encounter some of the greatest minds in human history. Regardless of whether they were discoverers or inventors, these fascinating lives are filled with countless memorable stories—stories filled with the same tragedy, triumph, and persistence of genius as that of the world's great writers, artists, and musicians.

Knowledge of Pythagoras, René Descartes, Carl Friedrich Gauss, Sophie Germain, Alan Turing, and others will help to lift mathematics off the page and out of the calculator, and into the minds and imaginations of readers. As mathematics becomes more and more ingrained in our day-to-day lives, awakening students to its history becomes especially important.

Sharon F. Doorasamy
Editor in chief

Editorial Consultant

In his youth, Curt Gabrielson was inspired by reading the biographies of dozens of great mathematicians and scientists. "I was driven to learn math when I was young, because math is the language of physical science," says Curt, who named his dog Archimedes. "I now know also that it stands alone, beautiful and mysterious." He learned the more practical side of mathematics growing up on his family's hog farm in Missouri, designing and building structures, fixing electrical systems and machines, and planning for the yearly crops.

After earning a BS in physics from MIT and working at the San Francisco Exploratorium for several years, Curt spent two years in China teaching English, science, and math, and two years in Timor-Leste, one of the world's newest democracies, helping to create the first physics department at the country's National University, as well as a national teacher-training program. In 1997, he spearheaded the Watsonville Science Workshop in northern California, which has earned him recognition from the U.S. Congress, the California State Assembly, and the national Association of Mexican American educators. Mathematics instruction at the Workshop includes games, puzzles, geometric construction, and abacuses.

Curt Gabrielson is the author of several journal articles, as well as the book *Stomp Rockets, Catapults, and Kaleidoscopes: 30+ Amazing Science Projects You Can Build for Less than $1.*

Ancient Mathematicians

one

Mathematics in Ancient Greece

Western culture owes a great debt to the civilization that flourished in ancient Greece. Some of the first democratic governments began in this Mediterranean society (although most Greek city-states were ruled by kings, who were often tyrants). Theater thrived in ancient Greece, and the works of such playwrights as Sophocles, Aeschylus, Euripides, and Aristophanes continue to resonate with modern audiences. The Greeks also produced beautiful physical art, and several later periods in art history hearkened back to the idealized figures of Greek sculpture. Architecture, music, poetry, philosophy—the development of all these disciplines also owes much to classical Greece.

In addition, the ancient Greeks were renowned for their advances in mathematics. This book will examine the lives and accomplishments of the leading lights of Greek mathematics from six century BCE to fifth century CE. This nine-hundred-year span includes two distinct eras in the history of Greek civilization: the Classical period and the Hellenistic period. The Classical period began in the early fifth century BCE, after the city-states on the Greek mainland fended off invasions by the mighty Persian Empire. With the threat of foreign domination removed, Greek civilization blossomed. During much of this period, the city-state of Athens was the foremost center of Greek culture and learning, though important artists and thinkers came from other places in the Greek world as well. Pythagoras, a philosopher and mathematician for whom the most recognizable theorem in geometry is named, was born on the island of Samos, near the coast of Turkey, but spent most of his life in the Greek colony of Croton in southern Italy. He is believed to have died near the beginning

A map of the ancient Greek world

of the Classical period. Most historians date the end of that period to 323 BCE, when Alexander the Great died.

As a youth, Alexander, a prince from Macedon, was tutored by the famous Greek philosopher Aristotle (384-322 BCE). After he succeeded his father on the Macedonian throne in 336, Alexander quickly brought all of Greece under his control. Thereafter he led his armies on a decadelong series of military campaigns, carving out a vast empire that extended south into Egypt and east through Persia and central Asia. Although Alexander's empire fragmented soon after his death, the Greek cultural ideals that he had brought to these far-flung lands endured. In fact, during the Hellenistic period—as the era following Alexander's death is called—the influence of Greek civilization continued to spread. During this period, Athens and other cities on the Greek mainland were eclipsed as centers of culture and learning by cities such as Pergamum in Turkey, Antioch in Syria, and Alexandria in Egypt.

Aristotle *(Courtesy of National Library of Medicine)*

Alexandria was the site of the renowned Museum, an institution of higher learning. Among its most famous teachers was Euclid—a mathematician who wrote the *Elements*, considered the founding work not only of geometry but of all logical thought. He lived during the early years of the Hellenistic period. Another celebrated Greek mathematician of the Hellenistic period, Archimedes, is thought to have studied at the Museum. But he spent most of his life

A marble carving of Euclid

Archimedes

in Syracuse, where he unlocked the secrets of the circle, laid the foundations for modern physics, and made important advances toward the discovery of calculus.

Historians date the end of the Hellenistic period to the first century CE, when Rome finished its conquest of Greece and the rest of the Mediterranean world. Nonetheless, remnants of Greek civilization endured under Roman rule in centers such as Alexandria. That city would, in fact, produce the last great mathematician of the ancient world: Hypatia, who was also renowned as a teacher and philosopher. Her death, in 415 CE, would come just six decades before the fall of the Roman Empire—an event that signaled the beginning of the West's long medieval period, during which much of the knowledge of the classical world was lost.

Besides mathematics, the one thing that links all of the people profiled in this book is education. Educational practices varied considerably throughout the Greek world. The Athenian model is perhaps the most famous, however. In

Athens, education began in the home. Children were taught by a parent or, in some households, an educated slave. Early instruction focused on music and gymnastics—an attempt to forge a balance of a fit body and an organized mind. Unlike in modern American math curricula, arithmetic and geometry were considered separate subjects.

In their twelfth year, Athenian boys would begin to attend school outside the home. There, they would learn grammar and basic logic and rhetoric, or public speaking. They would also learn more arithmetic—mostly calculations. In general, members of the merchant and artisan classes needed only to master computing skills for their everyday lives.

After several years in school, promising (and often wealthy) students would continue on to advanced education. Other students would leave school behind and enter the workforce. The advanced students would either attend an academy or be taught by a Sophist, a sort of freelance teacher of rhetoric. Students at an academy learned philosophy and advanced number theory. The focus now was on the science of numbers rather than their practical uses.

The academy learning environment was very different from the classroom system in use today. In modern classrooms, the teacher or professor often lectures to the class. Students take notes and demonstrate what they have learned on tests and by answering questions to which the teacher already knows the answers. This is learning that travels largely in one direction, from teacher to students. At the academy, classes were less structured. The teacher would stand in the center of a group of students, and they would ask him questions and talk to him about things that puzzled them. The teacher would try to answer these questions, and as he answered, a discussion would begin among all the students. The learning was directed by everyone, not just the teacher, although he could certainly push the conversation along certain lines. Socrates—probably the most famous teacher in ancient times—sought to lead students to the truth by asking question after question, letting the students come up with their own answers, which would either be accepted or discarded in favor of new explanations as learning continued.

Socrates

Three schools of special note arose at different times and places in the Greek world. The first was Pythagoras's school, which is believed to have been established in Croton around 518 BCE. It was Pythagoras who stressed music as a mathematical discipline rather than an entertaining pastime. Pythagoras's students were very devoted to him, living communally and taking vows of secrecy about certain matters. Eventually, his school was twice raided by the Croton government, causing many deaths and the destruction of buildings at the commune.

Around 387 BCE, Plato—one of Socrates' students—founded a school known as the Academy on the outskirts of Athens. Plato, who ranks among Western civilization's most influential philosophers, recognized the importance of geometry and mathematics. It is written that over the entrance to his Academy hung the inscription "Let no one ignorant of mathematics enter here." Unlike Pythagoras, who regarded contemplation of the mysteries of numbers as the highest end to an education, Plato considered mathematics the very basis for all intellectual thought. He said that mathematics should comprise the first ten years of any child's education. According to Plato, math was the best training for the mind; it encouraged logic and clear thinking. He also saw geometry in everyday life. Plato's Academy lasted more than nine hundred years. The Roman emperor Justinian closed it for good in 529 CE because he considered it a pagan institution.

The third great school of ancient Greece was Aristotle's Lyceum, founded in 335 BCE just outside Athens. Aristotle was famed for teaching as he walked through the grove in which the school was located. Eventually his students grew to be called the Peripatetics (patein means "to walk" in Greek; peri means "around"). Aristotle's school, while concerned with mathematics, broadened the course of study to natural sciences of all sorts.

This emphasis on mathematics throughout Greek education stands in contrast to the Greek numerical system, which was awkward to write and made quick calculations difficult. There actually were two systems of numbers in use. The first system, used in the first millennium BCE, was an additive model similar to the more familiar Roman numerals. The value of a number in an additive system has nothing to do with its position. There is a distinct symbol for each type of unit, and these are simply added to the figure to increase the number it represents.

Except for the number 1 (signified by a straight vertical line), every type of unit was symbolized by the first letter of its name. Number systems with this trait are called acrophonic. The Greeks had symbols for 5 and all multiples of 10 up to 10,000. The word for 5 was pente, so the symbol was a modified pi: Γ. (In fact, this symbol has a closer correspondence with the Greek letter gamma. This is because the gamma symbol originally stood for pi, and changed only after the symbols for these numbers were already solidly part of the culture.

Ten was deka, symbolized by delta (Δ). One hundred was hekaton, or an eta (H); 1,000 was called Khilioi, and took khi (X) as its symbol. And the topmost symbol was M for mu—the first letter of Murioi, meaning 10,000. Additionally, the Greeks used symbols for 50, 500, 5,000, and 50,000 by putting the ten-multiple symbol in subscript below the overhang of the pente symbol. The use of this five notation made the numbers more compact and easier to read and write by themselves. However, it made adding and subtracting very large numbers extremely difficult without the use of complex counting tables.

In the Greeks' additive number system, each unit making up a larger number appeared in the written version of that number. For example, counting from 1 to 10 would look like this: I, II, III, IIII, Γ, ΓI, ΓII, ΓIII, ΓIIII, Δ—every step adding a unit symbol until the next level of symbol was reached. There were no placeholders or zeroes. The number 103, for example, would be represented with an H (for hekaton, or 100) and three vertical lines for the ones: HIII. Since there were no tens, there was no notation at all between the hundreds and the ones places. This is part of what made calculation difficult—the Arab invention of the placeholder zero was a great advance in arithmetic. It makes all sorts of calculations, including such basics as addition, subtraction, multiplication, and division, much easier. The placeholder zero makes possible a positional number system, in which each digit represents by virtue of its location.

The acrophonic system was used for cardinal numbers, such as amounts of money or weights and measures. For a long time, ordinal numbers—first, second, third, and the like—had to be spelled out. Finally, in the fifth or fourth century BCE, a separate notation for ordinals was developed.

Complicating matters even further, Greek number notations varied from city-state to city-state —sometimes only in tiny ways, but sometimes dramatically. On the island of Cos, for example, no symbol for five was used (although the familiar symbols for 50, 500, and so forth were); thus, residents of Cos represented the number 9 with nine vertical slashes in a row. The version used in Athens was called the Attic system, and most local systems seem to be some variation of that. Each city-state had its own idiosyncrasies in its weights and measures, as well as its own monetary system.

Figures denoting sums of money had a slightly different notation than other numbers. The ones symbol would be replaced by the symbol for whatever unit of currency was being counted. If this notation were used in the United States, $561, for example, would be written 56$.

Greek money was not based on a decimal system; instead, money denominations were based on multiples of 6 and 8. The basic unit of Greek money was the drachma. There was one larger unit, the talent, which equaled 6,000 drachmas. The symbol for talent was a T; drachma was indicated by what looks like a plus sign without the left arm. There were also two smaller units: one drachma equaled six obols, and one obol equaled eight chalkos. Chalkos were indicated by an x, with a half obol symbolized by a C and a quarter-obol by a backward C; the obol itself was either indicated by an O or a straight vertical line.

While it might be difficult for an outsider to see how this money and number system worked, one piece of equipment brought it all together: the counting table. Functioning like an abacus, the counting table was used by merchants to do business. The most famous of these, the so-called table of Salamis, was discovered in 1846. With a row of symbols on either side and a set of divided horizontal lines running perpendicular to its length, it looks a little bit like a backgammon board, and it was initially thought to be a gaming table when first found. In fact, the symbols along the side are numerical symbols, indicating sums of money. These could be talents, drachmas, obols, or chalkos, depending on where the pebbles were set on the board.

In addition to the acrophonic system, the Greeks had a second system of numbers, called the alphabetical, or "learned," system. In this system, numbers correspond to the twenty-seven letters of the Greek alphabet. (The modern Greek alphabet only has twenty-four letters, but at the time this system was developed, there were three others that have since become obsolete: digamma, koppa, and san.)

The historians J. J. O'Connor and E. F. Robertson provide an excellent explanation of this system in their paper "Greek Number Systems." For the alphabetical system, the twenty-seven letters were divided into three groups of nine. The first group—alpha, beta, gamma, delta, epsilon, digamma, zeta, eta, and theta—represented the numbers 1 through 9. The second group—iota, kappa, lambda, mu, nu, ksi, omicron, pi, and koppa—represented the tens, 10 through 90. The hundreds, 100 through 900, were represented by the final column of nine letters: rho, sigma, tau, upsilon, phi, chi, psi, omega, and san. (Again, there was no symbol for zero.)

The benefits of this system are obvious. Numbers could be written much more concisely in the alphabetical system than in the acrophonic. To write 487 in the acrophonic system would take eleven characters, whereas in the alphabetical system it would take only three (delta pi psi, or $\Delta\pi\psi$). However, the alphabetical system posed two problems: If letters indicate numbers, how could numbers be written next to words? And how would numbers over 999 be designated?

The first problem was solved easily enough. In cases where numbers stood next to words and could be misinterpreted, it was customary to place a bar over the numbers. The second problem required a more elaborate solution, but the Greeks solved it as well. For thousands, they added a superscript (sometimes subscript) iota in front of the letters signifying 1 through 9, changing their meanings to 1,000 through 9,000. This brought them to 9,999.

For numbers 10,000 and higher, they introduced a new symbol: an M (for myriad) with a letter written in superscript directly above it. The small number over the myriad symbol signified the multiple of 10,000 the symbol represented. Therefore, an M with an alpha (A) above it signified 10,000; an M with a delta (Δ) over it signified 40,000; and an M with a sigma lambda epsilon ($\Sigma\Lambda E$) over it signified 2,350,000. If the number was even larger, and troublesome to fit over the M, it would sometimes be written in front of the M in superscript.

Two later Greek mathematicians developed notations for dealing with even larger numbers. Apollonius of Perga (ca. 262-190 BCE) developed a system that, instead of using multiples of the myriad to express large quantities, used powers of the myriad. For instance, an M with an iota over it would still equal 10,000, but an M with a beta over it would equal 10,0002, or 100,000,000. His contemporary Archimedes (ca. 287-212 BCE) developed what may have been a similar plan, except he used 100,000,000 as the standard value for M. He wrote about this system in his book *Sand Reckoner*, which has since been lost to history. It is claimed that he was able to notate up to 10^{64} or 10,000,000,000,000,000,000,000,000, 000,000,000,000,000,000,000,000,000,000,000,000,000.

These innovations notwithstanding, the Greeks' cumbersome number systems might be one reason why they considered mathematics and geometry two different disciplines—and why many Greek thinkers gravitated toward geometry. It, after all, had the advantage of being mostly about theory and ideal shapes, and not as much about the physical world.

In *These Were the Greeks*, H. D. Amos and A. G. P. Lang suggest a couple reasons for this. For one, Greek philosophers saw the senses as separate from the intellect, and they trusted the mind more than the senses. The senses could deceive. For example, when a straight branch is sticking halfway into a pool of water, the branch will look bent. This is because water refracts (or bends) the light coming from the part of the branch below the water while the light from the part above the water goes straight to the eye. To the Greeks, however, this was an example of the treachery of the senses. Also, they had not developed instruments to improve the senses, such as telescopes or microscopes.

Another reason for the divide between the intellectual and physical realms was that philosophers, by and large, were from the upper classes. They did not do physical labor, and they tended not to think about ways to make life easier for those who did. Because of this, many philosophers tended to confine their thought to matters of human nature. No experiment could reveal what "justice" or "duty" was; it was an exercise for the mind alone.

Geometry was in many ways the same. Despite the physical applications of the various theorems and propositions, geometry concerned itself primarily with perfect figures in an ideal imaginary space. As such, geometry was something of a community event among the educated upper classes. According to Geoffrey Lloyd, interviewed in Melvyn Bragg's *On Giants' Shoulders*, Greek mathematicians were driven by competition. One of them would propose a proof, and then the rest would try to find flaws in it. This would eventually result in a system of watertight proofs, exemplified by Euclid's methodical *Elements*. Lloyd contrasts this with Chinese society, which made great technological advances in roughly the same time. In China, there was only one person a scientist had to impress: the emperor. Of course, if a Chinese scientist turned out to be wrong in front of the emperor, his fate could be considerably worse than public humiliation among peers.

Without the work of countless anonymous people, nothing would be known today about what the ancient Greek mathematicians did or said. Up until the middle of the fifth century BCE, the Greeks favored the oral tradition, in which all information was memorized and passed down from generation to generation. Not until around 450 BCE did the Greeks begin writing history and other information on papyrus scrolls ten to fifteen feet long. This system had major drawbacks, however. In the damp Mediterranean environment, papyrus scrolls were prone to rotting away, even if stored untouched. When unrolled and read, they could easily rip or tear, and a careless reader could burn a manuscript when studying it by candlelight.

The only way to preserve these works was to copy and recopy them, time after time before the scrolls rotted. This was a painstaking and expensive process, and not all manuscripts were believed to be worth the trouble. In addition, with each successive copy, errors inevitably crept in. If the copying was done by someone unfamiliar with the field or the work, that person might unwittingly alter or omit crucial pieces of information. On the other hand, if the copier was an expert on the subject, he might take it upon himself to add facts or clarifications to the original text. Even when such information was correct, it still changed the original work, giving future readers the wrong idea about what was known when. While an expert copier's additions might have made the copy more useful at the time, they also made it less accurate as a historical document. Eventually there would be no original to consult, and the author's true intent might be lost.

Some figures of the time, such as Pythagoras and Socrates, did not write at all. In those cases, scholars must rely on their followers and students, as well as other people who wrote about them. Plato, for instance, wrote a great deal about Socrates in his *Dialogues*. But some of these secondhand accounts contradict each other, and others are just plain unbelievable. Scholars face the same difficulties in trying to examine the life of Pythagoras, as will be seen in the next chapter.

Establishing accurate dates for the various works of antiquity can also be problematic. Sometimes clues appear in the texts themselves. Many of the texts are accompanied by—or are part of—letters, which may mention contemporary events or the name of a reigning monarch. Some works have a dedication to a patron whose dates are known. Other texts include astronomical data, from which modern astronomers can extrapolate backward to a year when the stars and planets would appear as described. And this, in turn, can help scholars date people or works mentioned in the document—including works that have not survived to the present.

One other consideration scholars have when examining ancient works is the generation of copy they spring from. This is where the hand-copying system becomes a blessing instead of a curse. By comparing the errors in two similar texts, scholars can determine which of the two was copied first. If version one copies the errors of version two but has errors of its own, it is probably a copy of version two. If versions one and two each have unique errors as well as errors in common, it is likely that they were both copied from the same source. By studying the errors, historians can tell not only which was written first, but also which comes from the earliest source.

two
Pythagoras

For someone who lived so long ago, a remarkable amount is known about Pythagoras. The question is—how much of it is true?

Many ancient figures are known through their writing, but if Pythagoras wrote anything at all, nothing survives. Nonetheless, Pythagoras is one of the best-known philosophers to live before the time of Socrates, and his interest in mathematics popularized the discipline for centuries to come. The great British logician Bertrand Russell once called Pythagoras the "Albert Einstein of his day." His reputation in modern society is well entrenched; his name is practically synonymous with "mathematician." Whether that reputation is deserved is debatable, however.

A number of legends have sprung up about Pythagoras in the 2,500 years since he lived, and the scant surviving records of his time convey very little concrete evidence and

Pythagoras *(Courtesy of North Wind Picture Archives/Alamy)*

leave plenty of room for lies and distortion. Several biographies of Pythagoras were written in the third century CE, but that was eight hundred years after he lived. For reasons of their own, many writers of the time exaggerated Pythagoras's accomplishments. Historian Carl Huffman suggests that one biography, *On the Pythagorean Life* by Iamblichus, was an attempt to set Pythagoras up in the public's mind as a competing deity to the Christian Jesus. Not surprisingly, accuracy plays a distant second fiddle when elevating a man into a god. For this reason, "facts" about Pythagoras's life must be looked on with skepticism.

Pythagoras was born in Samos, a Greek island off the coast of Turkey, sometime around 570 BCE. His father, Mnesarchus, was a merchant from Tyre, a city in what is today Lebanon. According to legend, Mnesarchus settled on Samos after he brought corn there to quell a famine and was granted citizenship by the grateful people. He married Pythias, a native of the island, and they had three or four boys, one of whom was Pythagoras.

As children, Pythagoras and his brothers learned to play the lyre and recite poetry, including the works of Homer. Pythagoras often traveled on business trips with his father to Italy and Tyre, where he was taught by Chaldean and Syrian scholars.

Samos itself was not lacking in mathematical inspiration for the boy. The island was home to some great feats of engineering, the success of which was largely dependent on accurate mathematics. The island's most stunning achievement was the tunnel of Eupalinus, which was dug from both sides of a mountain at once. The two passages met in the middle with only slight correction. Samos was also located near Miletus, a city that was home to a group of philosophers known for their accomplishments in subjects such as mathematics and astronomy. Undoubtedly news of their work reached the island.

As Pythagoras was entering adulthood, his father died. Some sources say that Mnesarchus left his son with a letter of introduction to the philosopher Pherecydes, with whom Pythagoras studied for a few years on the island of Syros. Around the time he turned twenty, Pythagoras traveled to

Thales *(Courtesy of Visual Arts Library (London)/Alamy)*

Miletus to study with the elderly mathematician Thales and his student, Anaximander. Thales sparked Pythagoras's interest in math and gave him some advice that changed his life. He told Pythagoras that if he truly wished to learn the secrets of mathematics, he should go to Egypt.

Eventually, Pythagoras got his chance. Some sources indicate that he had become friends with Polycrates, the ruler of Samos. Others say he fled the king's tyrannical rule. One way or another, Pythagoras secured a letter of reference to Egyptian officials and eagerly left to further his studies.

Pythagoras was apparently the first Greek to learn to read Egyptian hieroglyphics. And at the temple at Diospolis, he learned the sacred rituals to become a priest in the Egyptian religion. Pythagoras would later incorporate many aspects of Egyptian religion into his teachings. Foremost among these was the value of secrecy.

Egyptian culture also showed Pythagoras much about mathematics and geometry. In Egypt, math was viewed in highly practical terms—as a tool for designing pyramids and other structures. The Egyptians already knew what would come to be known as the Pythagorean theorem—that the sum of the square of the lengths of the legs of a right triangle will equal the square of the length of the third side, called the hypotenuse. They had not proven it mathematically, but as a practical matter, they knew the formula worked. For them, that was good enough.

In 525, Persia invaded Egypt—ironically, with help from Polycrates at Samos. Sometime during the invasion, Pythagoras was taken prisoner by the Persians and brought to Babylon. During his time there, he apparently learned elements of Babylonian math. At some point, he either escaped or was freed; no records exist to say one way or the other. But in 522, the Persian ruler, Cambyses, died, as did Polycrates. Pythagoras was free to return to Samos. He did so around 520, when he was about fifty years old.

In Samos, Pythagoras founded a school called "the semicircle." However, his method of teaching, which was adopted from teachers in Egypt and which incorporated elements of

Egyptian spirituality, proved unpopular among the Greek students of Samos. Iamblichus, writing in the third century CE, claimed, "The Samians were not very keen on this method and treated him in a rude and improper manner." The poor reception of his ideas on his native island led Pythagoras to move to the city of Croton and start anew.

Croton (now called Crotone) was a city in southern Italy founded by Greeks around 710 BCE. Pythagoras's teachings were much more welcome there, and he soon attracted many students. Revered as a wise man, he led a society of loyal followers. Interestingly, he was held in such high esteem for reasons that had nothing to do with the mathematical accomplishments for which he is known today.

Chief among Pythagoras's spiritual teachings was a belief in reincarnation (or metempsychosis)—the view that, after death, a soul is reborn in another body. This belief stood in stark contrast to the Homeric view of death, which held sway at the time. It said that people's souls went to the underworld of Hades and there endured a dismal, hopeless existence. Pythagoreans believed that human souls could be reborn as other people, but also as animals. They may even have believed souls could be reborn as plants.

Xenophanes, a Greek philosopher of the late fifth and early fourth century BCE, told a story illustrating Pythagoras's belief in reincarnation. The story goes that Pythagoras once was walking along when he saw a man beating his dog. "Stop, don't keep beating it, for it is the soul of a friend of mine," Pythagoras told the man. "I recognize his voice." Xenophanes probably made the entire story up to ridicule Pythagoras. On the other hand, if the story is based on an actual incident, Pythagoras may simply have been jesting about recognizing his deceased friend's voice in order to end the beating of the dog.

Regardless of any friendship with dogs, various sources report that Pythagoras claimed he could remember his own soul's former lives. After living as a variety of plants and animals, Pythagoras eventually was reborn, he reportedly said, as a man named Hermotimus and then as a fisherman named Pyrrhus who lived on the island of Delos. After his stint as Pyrrhus, he was reborn as Pythagoras. Some writers recount that Pythagoras also claimed to be the Trojan hero Euphorbus and a prostitute named Alco. (*The Stanford Encyclopedia of Philosophy* suggests that again, this last part may have been meant in jest.)

The belief in reincarnation, however, leads to another tenet of Pythagoreanism: respect for all life. If any living thing—even an animal or plant—may have a human soul, then all living things must be accorded due moral consideration.

A group of Pythagoreans celebrating the rising of the sun.

Some writers have characterized Pythagoras as a vegetarian because of this belief. Although logic seems to point in that direction, the evidence is murky. Among Pythagoras's spiritual teachings were dietary restrictions for himself and his followers. One of the most prominent restrictions was on beans. Some scholars say this might be because they caused digestion problems, thereby making concentration difficult. Others suggest that Pythagoras thought beans somehow conveyed souls back into the world. This notion forms the basis of a legend recounted by Estelle DeLacy in her book *Euclid and Geometry.* According to the legend, Pythagoras was fleeing enemies intent on killing him when he came upon a bean field blocking his escape route. He refused to cross the field because doing so would destroy some of the plants, and his pursuers were thus able to catch and kill him. Contradicting this picture of a man unwilling to harm a bean even at the cost of his life is the account of Aristonexus, a student of Aristotle's. Aristonexus denied that Pythagoras had any qualms about eating beans. Rather, he wrote, "Pythagoras especially valued the bean among vegetables" and "especially made use of it."

Some scholars believe there are good reasons to doubt that Pythagoras was even a vegetarian. Another of the dietary rules ascribed to him prohibits eating the heart or womb of an animal. Presumably, other portions of the animal were considered fine to eat, or else why single out those organs? From Pythagorean prohibitions on eating certain species of fish, such as mullet and blacktail, it might similarly be inferred that the consumption of other kinds of fish was acceptable. It may be, as some writers have suggested, that only the higher tiers of Pythagorean society were vegetarians, so certain rules applied to some and not others.

All of this, however, perfectly illustrates how difficult it is today to be certain about any detail from Pythagoras's life. Only a century after his death, in fact, people were debating whether Pythagoras was a vegetarian or practiced animal sacrifice.

Aristotle recorded many of the maxims of the Pythagoreans. They came to be known as *acusmata* ("things heard") or *symbola* ("things to be interpreted"). Iamblichus divided these beliefs into three types of questions:

1) What is it?
2) What is best?
3) What must be done?

Questions of the "What is it?" variety tended to lead to strange notions. For instance, Pythagoras taught that the sound a gong makes when struck is a moaning of the spirit or demon living inside the brass. The second group of questions dealt in superlatives. The answer to "What is wisest?" would be "number," for example, and the answer to "What is loveliest?" would be "harmony." Questions of the third type, "What must be done?" would be answered with rules for living.

Pythagoras had rules for every aspect of life, not just for eating. These rules may have been meant to ensure a more advantageous reincarnation, although no records explicitly state this. For whatever reason, Pythagoreans did not act or dress like other members of society. Their rules for behavior call to mind religious observances of today, such as the Jewish practice of keeping kosher, the Muslim practice of bowing in the direction of Mecca during daily prayers, or the plain coats of Mennonites. Such practices need not always have religious significance. Members of the Free Masons wear an identifying ring, for example.

Some Pythagorean rules were about public conduct. Pythagoreans were easily recognized by their simple linen clothes, unadorned with decoration. Society members were not to urinate in public or sacrifice a white rooster, nor were they to use the public baths. Other rules had more to do with private behavior; Pythagoreans were encouraged to have children, to put their right shoes on first, and to reject material wealth. According to historian Walter Burkert, these kinds of rules for living were what made Pythagoras stand out from other philosophers. Many people would give rules of conduct for funerals, weddings, and the like, but Pythagoras focused on day-to-day life. After Pythagoras's death, people began to think that some of these rules were not meant to be taken literally, but were instead metaphors to be interpreted. Most evidence seems to support the literal view, however.

The one doctrine that Pythagoreans were best known for was that of silence. Some records say that new members of the Pythagorean society would remain silent for five years as a form of initiation. Others say simply that Pythagoreans were unusually taciturn in public debate, which gave their few public pronouncements more weight. But the code of silence most likely referred to the prohibition on discussing certain secret doctrines in public (and even, perhaps, with people at lower levels of Pythagorean society). Aristotle related that one of these secret doctrines claimed that there are three types of beings—gods, men, and those like Pythagoras.

It certainly seems that Pythagoras had a talent for self-aggrandizement. Rumors circulated that he had unearthly powers, signified by the golden birthmark on his thigh (said to be a sign that he was the son of the god Apollo). One marvelous story relates how a poisonous snake once attacked him, but he bit it to death. Another says that a river spoke to him, crying "Hail, Pythagoras!" as he was crossing it. Then there is the story of how a thief broke into Pythagoras's house, but was so terrified by the amazing things he saw that he fled, refusing to talk to anyone about his experience. (Just how this story could have spread, given the would-be thief's refusal to talk, is unclear.) Still other reports insist that Pythagoras appeared at both Croton and the town of Metapontum on the same day, at the same hour. Such stories almost certainly increased his fame—and attracted more followers.

Some contemporaries and near-contemporaries held Pythagoras in very high regard. The dramatist Ion of Chios (ca. 490-ca. 421 BCE) praised Pythagoras's knowledge of the soul after death. Many people did not consider his claims to be speculation, but true knowledge worthy of respect. The Athenian statesman and philosopher Empedocles (ca. 493-ca. 433 BCE) wrote that "whenever [Pythagoras] reached out with all his intellect he easily beheld all the things that are in ten or even twenty generations of men."

On the other hand, some people considered Pythagoras a fraud. The philosopher Heraclitus (ca. 540-ca. 480 BCE) was one such detractor. He called Pythagoras "the chief of swindlers."

These vastly divergent opinions of Pythagoras might be explained, at least in part, by his reputation as a performer of miracles. Among those inclined to believe the wondrous accounts, Pythagoras would have been highly esteemed. Skeptics, however, might have held the stories of miracles against Pythagoras—even if he had done nothing to perpetuate those stories.

Today, however, Pythagoras's "miracles" are easy to dismiss as the product of an earlier, more superstitious age. His importance, people of today would likely say, lies in his contributions to mathematics. Yet it is very possible—even probable—that Pythagoras actually had very little to do with the advancement of mathematics, aside from a pronounced interest in the subject.

One of the only early references to Pythagoras and mathematics of any kind are these two lines, penned by a writer named Apollodorus, about whom little is known but who may have lived in the fourth century BCE: "When Pythagoras found that famous diagram, in honor of which he offered a glorious sacrifice of oxen. . . ." The most remarkable thing about the verse at the time seems to be that it sparked an argument over whether Pythagoras would have sacrificed oxen if he were a vegetarian. But to modern eyes, the "famous diagram" is probably a reference to what has become known as the Pythagorean theorem.

The theorem, that the sum of the squares of the lengths of the legs of a right triangle equals the square of the length of the hypotenuse, is commonly denoted as $a^2 + b^2 = c^2$. Most people think that Pythagoras proved the theorem, but it in fact remained unproved until Euclid's time, centuries later. Failing that, Pythagoras is thought to have been the first person to notice the relationship between the legs of a right triangle and the hypotenuse, but this is not true either. The property had been well known among Babylonian and Egyptian mathematicians for centuries by the time of Pythagoras's early travels. It is possible that Pythagoras was the first person to bring this knowledge to Greece, but there are several other contenders for that distinction as well.

In all likelihood, the reason Pythagoras's name was attached to the theorem comes down to that ox sacrifice. Because he made such a big deal of the equation, he became associated with it in the public's mind.

Pythagoras is also associated with another theorem, called the angle sum theorem. This theorem states that the three angles of a triangle will add up to 180 degrees. (This was originally stated as "the sum of two right angles," since the degree system of measure had not been invented yet.) Again, Pythagoras did not prove this principle, but he might have discovered it.

While Pythagoras may not have been the mathematical innovator he is reputed to be, he likely did have a fascination with numbers. Someone influential in the Pythagorean society certainly did. The Greek word *mathematikos* originally referred to learning of all sorts, but the Pythagoreans so emphasized the disciplines today known as math that the word shed its wider meaning. (It still lives on in words like *polymath*, which is a person who knows many things in different subjects.)

The designations of "odd" and "even" numbers come from the Pythagorean tradition. Pythagoras and his followers also ascribed other characteristics to numbers, including masculinity and femininity (odds were masculine, evens feminine), reason (one), opinion (two), and justice (four). Ten was considered the best number, since it was the sum of the first four numbers, called the *tetraktys.*

The tetraktys might have been so special because it represented the number of points needed to define the universe. The first two points make a line, a one-dimensional figure. Add a third point, and the line becomes a two-dimensional triangle. A fourth point, if not on the same plane as the other three, makes that triangle solid—a pyramid, to be exact. Four nonplanar points thus define the three-dimensional world.

Pythagoreans were interested in mathematics not simply as a practical craft, but as a science that existed purely in the mind. They had a saying: "A figure and a platform, not a figure and sixpence." They disregarded the material and practical uses for math; Pythagoreans were more interested in each new bit of information as a platform for further learning than as a way to make money. (Euclid would later exemplify this attitude with his *Elements,* methodically building an entire system of geometry from a handful of assumptions.)

Pythagoreans believed that the universe was mathematical. All things, they said, could be expressed in a ratio; everything was literally rational, and could therefore be understood.

This belief in the rationality of the universe was severely shaken by a discovery that was so shocking to the Pythagoreans that they did their best to keep it secret from all but their inner circle. Using what became known as the Pythagorean theorem, Pythagorean mathematicians tried to assign a value to the diagonal of a square. The diagonal cuts the square into two identical right triangles, and the diagonal is the hypotenuse of those triangles. So it seemed likely that the mathematicians, knowing the lengths of both legs of the triangle, could compute the length of the third side.

But when they tried, they found a number that could not be expressed as a ratio (that is, as a fraction). Since all sides of a square are equal, it does not matter what the number assigned to them is—it is one unit. Therefore, the lengths of both legs of the right triangle made by bisecting the square along its diagonal are also one unit. When these numbers are plugged into the Pythagorean equation, the result is as follows: $1^2 + 1^2 = x^2$, or $x^2 = 2$. A reasonable approximation of this number is 1.41421356. But the number does not stop in the hundred-millionths place—it keeps going, without ending or repeating. It cannot be expressed as a fraction, which is the best that the Greek number system was able to do. Because there is no ratio that can express it, it is called an irrational number. Pythagoreans also called it an *algon,* which literally means "not to be spoken." The existence of irrational numbers was the greatest secret of the Pythagoreans, because it undermined their strong belief that everything in the universe could be expressed as a rational number.

Some Pythagoreans spoke of it anyway. Sometime after Pythagoras's death, his student Hippasus disclosed the existence of the irrational number. He was soon murdered, thrown overboard from a boat. This very well could have been an assassination to keep the square root of 2 a secret.

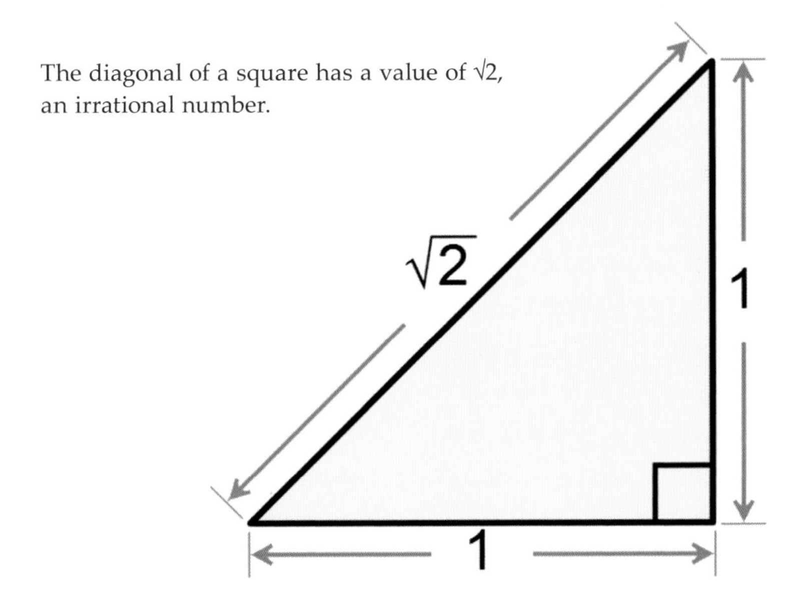

The diagonal of a square has a value of $\sqrt{2}$, an irrational number.

How murder squares the Pythagorean idea of respecting all life is anyone's guess.

Pythagoras's interest in mathematical ratios also found expression in music. Legend says that one day, while walking past some blacksmiths, Pythagoras noticed that each smith's hammer rang with a different tone. Upon further study, he discovered it was the weight of the hammer that made the sounds different. Some of the tones sounded good together; others were dissonant. The story goes that by examining the hammers, Pythagoras ascertained the ratios that became the founding intervals of music.

This story is almost certainly false. For one thing, the metal a blacksmith hammers is more likely to ring than the hammer itself. Nonetheless, the intervals of music do come from the Pythagorean tradition. And, strangely, they are composed of ratios of the four numbers of the tetraktys. A ratio of 1:2 makes an octave. That is, a guitar string one foot in length will make a sound one octave above a two-foot-long string of the same material. A ratio of 3:2 produces a fifth, and 4:3 makes a fourth. These results hold true for a single string fingered at different points. A two-foot-long string will produce a sound that is one octave lower than the sound produced if that string is held at one foot and plucked. And that same string held at six inches will produce a sound two octaves higher than the original tone. At three inches, it will be an octave higher still. Musical instruments still make use of this principle today.

Another Pythagorean geometric feat was constructing regular solids. These solids were considered extremely important. To be a regular solid, a construction must have sides made of a regular polygon—a triangle, square, or pen-

Pythagoras playing a stringed instrument *(Courtesy of Lebrecht Music and Arts Photo Library/Alamy)*

tagon, for example. (Regular also means that the sides and angles of the polygons are equal—an equilateral triangle is a regular polygon, for example, but an isosceles triangle is not.) Pythagoras or his followers discovered that there are only five regular solids: a pyramid (created with four triangles, with one at the base); a cube (made with six squares); an octahedron (an eight-sided solid with faces of equilateral triangles); a dodecahedron (a twelve-sided form with regular pentagons as faces); and an icosahedron (a twenty-sided form using equilateral triangles as bases). It is said that Pythagoras was able to construct the first three solids during his lifetime; his followers continued the effort, finding the last two. These solids are now called the platonic solids. A platonic solid is a polyhedron all of whose faces are congruent regular polygons and where the same number of faces meets at every vertex. The best known example is a cube (or hexahedron) whose faces are six congruent squares.

Pythagoras was said to have had a long-standing policy about keeping out of politics. Yet around 510 BCE, a situation arose in which he apparently considered it his duty to intervene. Some Pythagoreans had traveled to Sybaris, a Greek city in southern Italy that was ruled by a tyrant. There the travelers were captured and killed. A while later, a group of Sybarites arrived in Croton as refugees. Pythagoras urged his city to show mercy and let the Sybarites stay. City officials did as he wished, which sparked a war with Sybaris. Croton eventually won that war, but its start earned Pythagoras enemies in the government.

Those enemies exerted their influence, causing a raid on the Pythagorean commune around 500. Buildings were set ablaze and many Pythagoreans were killed. Pythagoras himself fled for his life. (This is the point where the legend says he refused to cross a bean field and was killed by his pursuers, but that story is almost certainly false.) It is unclear what happened after the raid. Most sources say that Pythagoras fled to Metapontum. Some claim he committed suicide soon afterward. Others, however, say he lived in Metapontum until his death in 470. Whatever the circumstances of his death, this date seems a bit too convenient to be plausible. In 470, Pythagoras would have been one hundred years old had he been born in 570, as was believed. Given that Pythagoreans viewed 10 as the perfect number (since it is the sum of the four numbers of the tetrakys), 10 squared, or 100, might have been considered the "perfect" age for Pythagoras to die. Those who spread this story were probably more concerned with building a legend than reporting the facts. More reliable accounts suggest that Pythagoras died in Metapontum around 490.

Regardless of when Pythagoras died, the Pythagoreans continued on without him. But so did his enemies. Some fifty years after the first attack on the Pythagorean community, there was another attack, resulting in more deaths and burned buildings. Nonetheless, the Pythagoreans persevered.

After his death, Pythagoreans split into two groups: the *acusmatici* and the *mathematici.* The *acusmatici* were devoted to following the Pythagorean maxims. Everyone, including the *mathematici,* recognized them as genuine Pythagoreans. Not so with the *mathematici.* They were closer in philosophy to Hippasus, the man who was thrown overboard for revealing the square root of 2. They claimed that Pythagoras had taught everyone his maxims, which of course were worth study. But they said he taught his way of thinking to his youngest students, explaining the rationale behind his teachings. Instead of blindly following the rules made in the past, the *mathematici* claimed, they were applying Pythagorean teachings to the current situation—and sometimes getting different results. Because of this, the *acusmatici* thought they were frivolously pretending to be Pythagoreans. In short, they thought the *mathematici* were poseurs.

The first known book written by a Pythagorean was by Philolaus of Croton (ca. 470-ca. 390 BCE). His work was the basis for Aristotle's later writing on Pythagoras. Philolaus also brought a few new ideas to Pythagoreanism. The first was a way to look at the world as made up of "limiters" and "unlimiteds." The latter category embraced the materials of which objects are composed: stone, wood, water, and so on. A limiter described the material's boundaries—for instance, the shape of a mountain, or a tree, or a river. A simple analogy, albeit in two rather than three dimensions, might be a child's coloring book. The unlimited is the color, and the limiter is the outline to color within.

Philolaus was also concerned with cosmology. He knew that the earth is a planet (he was perhaps the first to reach this conclusion). But he believed that the earth circled a central fire, as did the five other planets visible to the naked eye and known to the ancients, the sun, the moon, and the stars. Because ten is the perfect number and he only had nine elements circling the fire, Philolaus added another planet, called "counter-Earth." The reason counter-Earth could not be seen, he said, was that it was always on the other side of the central fire.

Archytas of Tarentum, who died in the mid-fourth century BCE, was probably the last prominent Pythagorean. A contemporary of Plato, he proved what Pythagoras thought about music ratios: that the octave, the fifth, and the fourth could not be divided in half. He is also credited with founding the discipline of mathematical mechanics.

Some of Plato's followers would write that certain of his ideas had their roots in Pythagorean thought. In many cases, this was not so. However, because Pythagoras was beginning to be revered as a semi-divine figure by Plato's time, claiming that Plato's ideas were based on those of

Archytas of Tarentum *(Courtesy of Mary Evans Picture Library/Alamy)*

Pythagoras was a way to give them credibility. Yet the only reference to Pythagoras in Plato's work, which is found in *The Republic*, portrays him not as a mathematician but as a beloved teacher who showed how best to live in a manner pleasing to the gods.

As time went on, the legend of Pythagoras grew greater than the man himself. Around 200 to 300 CE, seven or eight centuries after his death, books supposedly written by Pythagoras appeared. Scholars in the fifteenth century discovered that the books—as well as seventy-one lines of the *Golden Verses*, a work that had also been attributed to Pythagoras—were forgeries. What was claimed to be Pythagoras's writing was only the result of so much smoke and mirrors. As for the man himself, it is difficult to say.

The Shapes of Numbers

Among his other numerological interests, Pythagoras was also interested in the shapes of numbers. The name "square number" is not a coincidence: when represented by pebbles or dots, square numbers such as 4, 9, and 16 can be arranged to form a square. Sixteen pebbles form a four-by-four square; twenty-five pebbles comprise a five-by-five square.

Pythagoras (or one of his followers) noted that successive squares have an interesting property. Every square is the sum of consecutive odd integers, starting at 1. For instance, 4, the second square number, is the sum of the first two odd integers, 1 and 3. Nine is the sum of the first three odd numbers (1+3+5); the fourth square, 16, is the sum of the first four odd numbers (1+3+5+7), and the pattern repeats. (One, the first square number, is the sum of itself.)

Square numbers were not the only number shapes that Pythagoras recognized. He also studied "triangular" numbers—numbers that in pebble representation could form a triangle. These were 1, 3, 6, 10, and so on. A ten triangle has four pebbles at its base, then three above it, then two, then one at the tip. Triangular numbers are formed by the sum of all integers, odd and even. The fifth triangular number, 15, is formed by adding the first five numbers: 1+2+3+4+5.

Pythagoras also studied sums of even numbers. These could be arranged into rectangles slightly longer on one side than the other. Therefore, they were called "oblong" numbers. Two pebbles plus four pebbles would form a rectangle of six pebbles, arranged two by three. Add another six (the third even number) and the result would be a three-by-four rectangle, making 12, the third oblong number. As interesting as these patterns may be, triangular numbers and oblong numbers never proved as useful as square numbers, and eventually they fell out of favor.

Timeline

ca. 570 BCE Born in Samos, a Greek island.
ca. 535 BCE Travels to Egypt.

ca. 525 BCE Persia invades Egypt; Pythagoras is captured.

ca. 520 BCE Returns to Samos; he soon moves to Croton.

ca. 510 BCE Urges Croton to offer Sybarites asylum; decision to grant asylum starts war with Sybaris.

ca. 500 BCE Pythagorean school raided; Pythagoras flees.

ca. 490 BCE Dies.
ca. 450 BCE Second raid on Pythagorean school.

three
Euclid

If many of the details of Pythagoras's life are uncertain, virtually no biographical information exists for Euclid of Alexandria. It is not a matter of knowing which texts are trustworthy; rather, when people wrote about Euclid, they wrote about his many books, particularly the *Elements.* Euclid's life story took a backseat to his work.

Most of what is known about Euclid's life comes from extrapolation. He was one of the first (and greatest) teachers at the famed Museum of Alexandria. In all likelihood, he was there at or shortly after the Museum's founding, in the early third century BCE, by Ptolemy I Soter. This puts his birth and death dates between the death of Plato (approximately 347 BCE) and the emergence of Archimedes (in the middle of the third century BCE). It is possible that, before leaving for the Egyptian coastal city of Alexandria, Euclid also taught at Plato's Academy.

Euclid

Beyond that, most of what is known of the man is his bibliography. Euclid wrote at least ten treatises. Only six of them still survive, in one form or another, although references to several more are found in other writings.

One of the lost works was a four-volume treatise called *Conics.* Archimedes referred to it in his own work. From his description, *Conics* was most likely a compilation of all the knowledge to that time on curves and conic sections. Conic sections are segments of a cone. For instance, if a cone is cut parallel to its base, the shape of the new face is a circle. But if the cone is cut at an angle to its base, the resulting face will be an ellipse.

Another of Euclid's lost works, titled *Porisms,* is said to have been composed of three volumes. Scholars are not sure what the subject of this work was. One possibility is that it was a start to a theory of geometric loci. (Geometric loci are series of points that can be defined by an equation. For example, a circle is a series of points such that all points are a certain distance—the radius, or r—from the center.) If this is the case, the Greeks may have stopped reproducing *Porisms* because they had mastered and moved beyond the theory, and obsolete books were not worth the expense or effort of copying.

Euclid is also credited with a two-volume work called *Surface Loci.* This may have been supplanted by Archimedes' own work on the subject.

The most intriguing of Euclid's lost works is called *Pseudaria*, or *Fallacies.* In it, historians suspect, Euclid attempted to teach students what *not* to do when constructing geometrical figures and attempting proofs. In other words, the book was likely an effort to steer readers away from logical fallacies and the dead ends that students of geometry are prone to encounter.

A handful of other works once bore Euclid's name but have since been revealed as misattributed. These include a volume called *Mechanics,* another called *Catoptrica,* and a book now known to be by the second-century CE Greek theorist Cleonides, called *Introduction to Harmony.* Another work, *Sectio canonis* (The Division of the Scale), was very similar to Pythagorean music theory. This may or may not be the work of Euclid; some historians consider it part of a larger, lost work called *The Elements of Music,* which Proclus—a Greek philosopher of the fifth century CE—attributed to Euclid.

Among Euclid's surviving mathematical works are *The Data,* a collection of ninety-four advanced geometric propositions, and *On Divisions of Figures,* which deals with the outcome of dividing geometric figures with one or more straight lines. The latter was reconstructed from existing Arabic and Latin translations; no Greek original for *On Divisions of Figures* has ever been found.

Euclid's surviving works also include two books on scientific rather than purely mathematical subjects. *Optics* is an examination of perspective. *The Phaenomena* provides an introduction to astronomy, though it becomes more of an exercise in spherical geometry, as Euclid makes no attempt to explain the movement of the sun, moon, or planets. Rather, he considers the universe as a collection of circles and proceeds to discuss the figures and their properties with only passing reference to the heavens.

Euclid's enduring fame, however, rests with his seminal treatise on geometry, the *Elements.* It provided the framework for almost all work in geometry since his time.

Here and on page 57 are pages from a tenth-century copy of Euclid's *Elements*.

Interestingly, there is no evidence that Euclid himself ever used the term *geometry* in reference to the *Elements*. From the Greek roots *geo* (meaning "earth") and *metrein* ("measure"), the word literally meant to measure the earth. Originally it was used to signify land surveying. *Stoicheia*, or "elements," was the word Euclid chose for the title of his landmark treatise. The ancient Greeks believed that four elements—earth, air, fire, and water—were the building blocks of the entire world, and Euclid probably wanted to indicate that his work similarly underpinned the subject we now know as geometry. His confidence proved justified: Proclus was not exaggerating when he declared that the *Elements* is to the rest of mathematics what the letters of the alphabet are to language itself. Today Euclid's work is still considered the foundation of geometry.

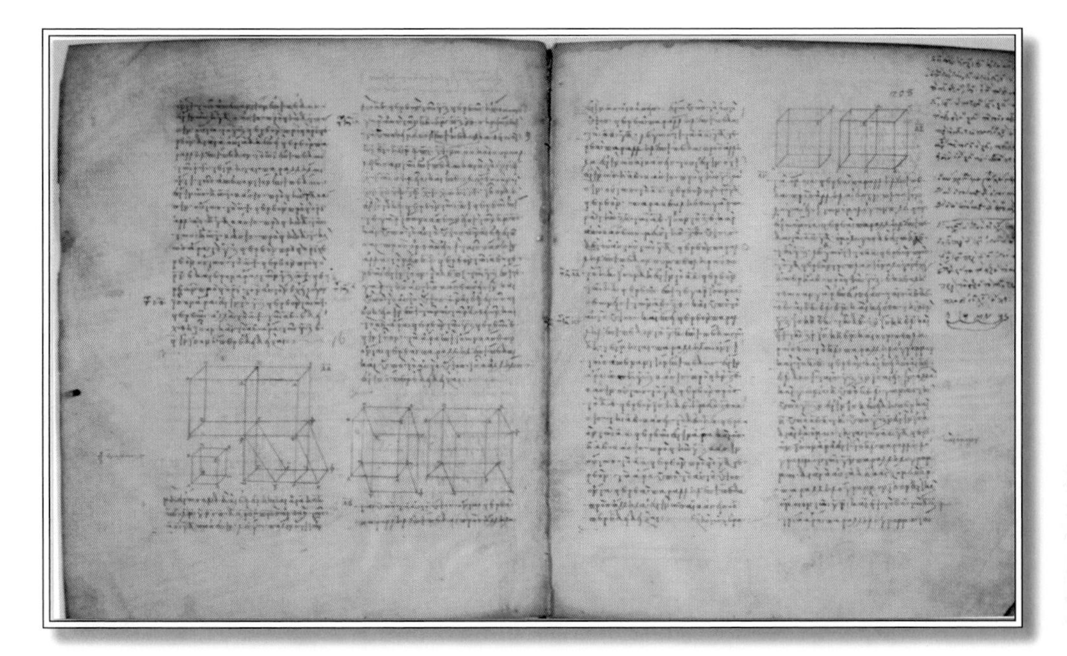

Euclid was certainly not the first person to think and write about geometry. Before the *Elements*, in fact, at least four other ancient works compiled previous geometrical treatises. These four editions have been lost—in large part because Euclid's work rendered them obsolete. The *Elements* was so successful that it proved to be the last compilation of ancient geometric knowledge. No one managed to outdo Euclid—if anyone even tried.

Among the mathematicians whose work Euclid drew on for the *Elements* were Hippocrates of Chios (not to be confused with Hippocrates of Cos, for whom the doctor's oath is named) and Theudius, a previous compiler of geometrical knowledge. But while Euclid compiled the work of prior geometers, he added something immeasurably important. What made the *Elements* stand out from all works before it, and made it such a model for the works to follow, was its structure.

Aristotle, whose life span probably overlapped Euclid's, had strong ideas about scientific and mathematical inquiry. Euclid's structure epitomizes one of these Aristotelian ideals: that the most trustworthy proofs are those with the fewest underlying assumptions. The less taken for granted at the start of the process, the more the result is shielded from errors stemming from a faulty premise.

Euclid begins the *Elements* by making only ten assumptions, five about geometry in particular and five about science and the world in general. The five geometric assumptions are called postulates. Euclid calls the five more general assumptions "common notions." They have also taken on the name "axioms."

Along with the postulates and axioms, Euclid is very careful to define his terms. These early definitions cover a variety of geometric terms: a point ("that which has no part"), a line ("a breadthless length"), a circle ("a plane figure contained by one line such that all straight lines falling upon it from one point among those lying within the figure—the center—are equal to one another"), and so on.

Euclid's five postulates—his assumptions about geometry—are, for the most part, fairly basic. They are intrinsic to the common understanding of space, and without them (or at least four of them) geometry might not even be possible. The postulates are:

Postulate 1: It is possible "to draw a straight line from any point to any point."

Postulate 2: It is possible "to produce a finite straight line continuously in a straight line."

Postulate 3: It is possible "to describe a circle with any center and radius."

Postulate 4: "All right angles equal one another."

While right angles were not yet measured as "90 degrees"—that unit had not yet been invented—this signaled that they were uniform. Euclid had defined right angles in definition 10: "When a straight line standing on a straight line makes the adjacent angles equal to one another, each of the equal angles is right, and the straight line standing on the other is called a perpendicular to that on which it stands." What Euclid is saying with this postulate is not just that the two angles on either side of the perpendicular are equal, but that all right angles, anywhere, are equal to each other.

> **Postulate 5:** "If a straight line falling on two straight lines makes the interior angles on the same side less than two right angles, the two straight lines, if produced indefinitely, meet on that side on which are the angles less than the two right angles."

The wording of the fifth postulate is somewhat difficult to understand, but Euclid is describing lines that are not parallel. These lines, he says, will eventually meet. When a third line crosses two non-parallel lines, the sum of the interior angles on the side on which they meet will measure less than 180 degrees, or "two right angles."

This last postulate, known as the parallel postulate, is the only one of the five that has met with any serious contention. In fact, Euclid himself did not seem entirely comfortable making the assumption. It was not until his twenty-ninth proposition that he actually resorted to relying on proposition 5 for a proof. (Eventually, branches of geometry that do not rely on this postulate were explored. These were dubbed non-Euclidean geometry.)

Non-Euclidean Geometry

The postulates Euclid uses at the start of the *Elements* are assumptions necessary to begin the work of proving the subsequent propositions. They are not proven themselves; rather, their truth is taken as a given.

Of the five postulates with which Euclid opens the *Elements*, four are absolutely essential to the work. They are used throughout the early proofs, upon which everything else rests.

While also important, the fifth postulate—known as the "parallel postulate"—is not quite as crucial to Euclid's work. The fifth postulate says that through any given point, it is possible to draw one and only one line parallel to a given line. Euclid avoids using this postulate for as long as possible, relying on it at last to prove proposition 29 in Book I. A sizeable chunk of the *Elements* is founded on this idea, but a considerable number of propositions have been proven without its use. And thanks to Euclid's meticulous structure, it is easy to identify both the propositions that depend on the parallel postulate and those that stand without it.

In the nineteenth century, various mathematicians began to explore what geometry would be like if this postulate were not taken as a given, and in fact if its opposite were true: that through any given point, it *is* possible to draw more than one line parallel to any given line. This certainly seems to run counter to the physical evidence. If a piece of paper represents a plane, it looks pretty obvious that parallel lines drawn on it will not touch. But if that paper is folded or crumpled, the lines touch just the same.

Some of the people who explored this concept were Nicolai Lobachevsky (1792–1856), János Bolyai (1802–1860), Georg Friedrich Riemann (1826–1866), and Albert Einstein (1879–1955). The first to do so was probably Carl Friedrich Gauss (1777–1855), although he kept his study on it a secret until Bolyai and Lobachevsky began publishing.

Gauss

Three types of non-Euclidean geometry developed: hyperbolic, elliptical, and absolute. In hyperbolic geometry the parallel postulate is false, because two lines parallel to a given line can pass through the same point. The parallel postulate is also false in elliptical geometry, because there are some points through which *no* line can be drawn that is parallel to a given line. In absolute geometry, neither alternative replaces Euclid's parallel postulate; rather, absolute geometry is simply geometry based on the remaining four postulates. All three of these geometries, which require a radical reinterpretation of the properties of the universe, are difficult to explore with the tools available during Euclid's time.

Euclid's axioms, or common notions, are a bit easier to understand. They apply to all sciences, and have broader reach than geometry.

> **Common notion 1:** "Things which equal the same thing also equal one another."
>
> **Common notion 2:** "If equals are added to equals, then the wholes are equal."
>
> **Common notion 3:** "If equals are subtracted from equals, then the remainders are equal."
>
> **Common notion 4:** "Things which coincide with one another equal one another."

(By "coincide," Euclid was referring to figures that matched perfectly in sides, angles, and so on if one was overlaid on the other.)

> **Common notion 5:** "The whole is greater than the part."

Euclid combined these notions with the postulates to prove what he called propositions. Essentially, the propositions stated that, using only what was already known, a certain assertion could be proved. Throughout Euclid's *Elements,* each proposition would go through these four basic steps: the statement of the proposition; the construction of the figure in question; a demonstration that the figure meets the requirements; and a conclusion that the proposition has been proved.

For instance, proposition 1 is: "To construct an equilateral triangle on a given finite straight line." Euclid does this by first drawing a line segment, *AB*. Then he constructs two circles around the segment. The first has point *A* as its center and point *B* as a point on the circle (therefore, segment *AB* is its radius). The second has the opposite construction: point *B* is its center, and point *A* is on the circle itself. Segment *BA* (which is the same as segment *AB*) is the radius.

These two circles will intersect above segment *AB* at the point *C*. (The circles also intersect at a second point below *AB*, but that has no bearing on the proof.) A line segment drawn from *A* to *C* would be equal to segment *AB*, since both *C* and *B* are points on circle 1, and therefore must be the same distance from its center, point *A*. Similarly, a line segment drawn from *B* to *C* (completing the triangle *ABC*) must be the same length as segment *AB*, since both *A* and *C* are points on circle 2, and therefore must be equidistant from its center, point *B*. And, using common notion 1, if segment *AC* is equal to segment *AB*, and segment *BC* is equal to segment *AB*, then segment *AC* must also be equal to segment *BC*. And since all three sides of triangle *ABC* are equal, it is by definition an equilateral triangle, proving the proposition.

Of course, as nice as it is to have an equilateral triangle, constructing the figure was only the first step in the hundreds of propositions Euclid proved. But the triangle was an invaluable tool along the way—as were most propositions—because once a proposition was proved, it could be accepted as true for use in further proofs. So Euclid was able to construct an equilateral triangle for use in proving proposition 2 (that two lines of equal length could be constructed from the same point), and he used proposition 2 to prove proposition 3 (that a segment could be cut from a longer line such that it would be equal in length to a shorter line). And on it went, with Euclid building each proof from the last, accumulating an increasing supply of geometrical tools as he did so.

It was this structure that made Euclid's *Elements* so sturdy, so impervious to criticism, and so immune from being supplanted by new ideas. Each postulate was meticulously proven and backed up by its predecessors. Even better, if one postulate did somehow turn out to be faulty, Euclid's construction made it easy to see which of the later postulates would also be called into question (and either need to be proved again or discarded), and which could still stand despite the fracture in the system. "Euclid is not the father of geometry," writes Petr Beckmann in his *A History of Pi*. "He is the father of mathematical rigor."

In the whole of Book I, Euclid proceeded to prove forty-eight propositions. These included proving that if a triangle has two equal sides, then the angles opposite the equal sides must also be equal (proposition 6); that if two straight lines cross one another, the angles at opposite sides are equal (proposition 15); and that if two triangles have two angles equal to each other, as well as one side equal in length, then the two triangles are equal (proposition 26). This last became known as the angle-angle-side proposition, or AAS. Later, he proved that the three angles in any triangle are equal to two right angles, or 180 degrees (proposition 32).

During the Middle Ages, one of Euclid's earliest proofs earned a colorful nickname because of its difficulty. Teachers dubbed proposition 5—that the base angles of an isosceles triangle are equal—the *Pons Asinorum,* or "Bridge of Asses," since it thwarted so many of their students, who often could not understand the proposition or even see the need for the proof. A name that was undoubtedly more agreeable to students appears in the writings of the English monk and philosopher Roger Bacon around 1250 CE. *Elefuga,* from Greek words meaning "escape from misery," apparently derives from the fact that Euclid's fifth proposition was the last proof students of the day needed to know before they could put geometry aside and move on to other studies.

Book I ended with Euclid's proof of the Pythagorean theorem—that the square of the side of a triangle opposite a right angle (the hypotenuse) is equal to the sum of the squares of the two remaining sides. Furthermore, he followed that up by proving the converse of the theorem—that in any triangle, if the square of one side equals the sum of the squares of the other two sides, the angle opposite the first side is a right angle.

Euclid's methodical approach required considerable thought and effort. But he considered the steps of the proof every bit as important as the conclusions of the propositions. Proclus wrote that Ptolemy I, the king of Egypt and the founder of the Museum of Alexandria, once asked if there might be some shorter way to learn the *Elements*. Euclid's answer was telling. According to Proclus, the mathematician told the king that "there was no royal road to geometry." Kings and commoners alike must take the same steps to understand the discipline.

Books II through IV of Euclid's *Elements* continued Book I's focus on plane geometry. Book II concentrated on geometric algebra; Book III dealt with circles; and Book IV covered regular polygons, including the pentagon.

Books V and VI dealt with ratios. Book V considered ratios and proportions on their own terms, while Book VI tied them back into plane geometry. This is a good example of how the books themselves mirror the building-block structure of the propositions; Book VI would have not worked without the foundations of Books I through IV and Book V.

Books VII through IX covered number theory (of positive integers greater than 1). Book VII begins with twenty-two

Euclid presents his *Elements* to Ptolemy I Soter. *(Courtesy of Mary Evans Picture Library/Alamy)*

addition definitions, this time concerned with numbers rather than geometric figures. They include unit ("that by virtue of which each of the things that exist is called one"); number ("a multitude composed of units"); even number ("that which is divisible into two equal parts"); odd number ("that which is not divisible into two equal parts, or that which differs by a unit from an even number"); and prime number "that which is measured by a unit alone"). In Euclid's usage, "measured by" means "divisible by," so a prime is a number divisible only by 1 (and the number itself). His definition of multiple ("the greater number is a *multiple* of the less when it is measured by the less") means that a larger number is a multiple of a smaller number that is divisible into it, in the way that 12 is a multiple of 4.

Book VII concerned itself with finding the greatest common divisor between any two numbers, Book VIII examined geometric sequences of numbers, and Book IX worked on proving that there are an infinite number of prime numbers. Book X brought the focus back to plane geometry, creating classifications for lines and arcs.

Books XI through XIII dealt with the construction of three-dimensional figures (called *stereometria* in Greek). Book XI dealt with the intersection of planes and lines through certain solids. Book XII examined the relationships between figures such as circles, squares, triangles, pyramids, cones, and cylinders and proved that the area of a circle is a function of the square of its diameter. Book XIII constructed the five platonic solids: pyramid, cube, octahedron, dodecahedron, and icosohedron. (Originally the platonic solids and their regularities were discovered by the Pythagoreans and were initially called the Pythagorean

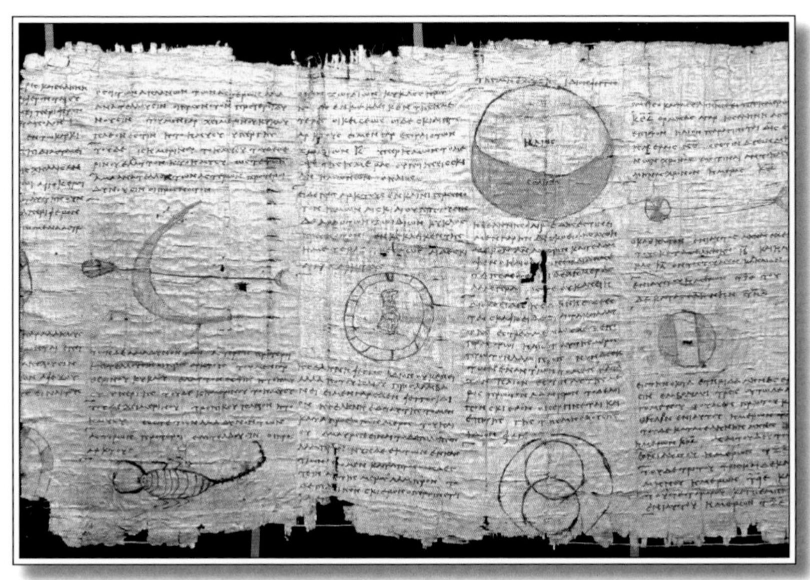

Eudoxus of Cnidus's writings about astronomy on an ancient papyrus scroll. *(Courtesy of Réunion des Musées Nationaux/Art Resource)*

solids. The Greek philosopher Plato described the solids in detail later in his book "Timaeus.")

For Book XII's work with the circle, Euclid used a method invented by an earlier mathematician, Eudoxus of Cnidus. Eudoxus, who lived during the first half of the fifth century BCE, is notable for three reasons. First, he formulated an explanation of the movement of heavenly bodies that survived until Copernicus in the sixteenth century.

Second, his theory of proportion—which explained and allowed for irrational numbers—resolved the so-called logical scandal in Greek mathematics precipitated by the Pythagoreans' discovery that the diagonal of a square with sides of one unit could not be expressed as the ratio of natural numbers. Euclid would deal with Eudoxus's theory of proportion in Book V of the *Elements*.

Third and most important, Eudoxus developed the method

of exhaustion, a way to find areas and volumes. This was the method Euclid used to prove that the area of a circle is a function of the square of its diameter. Even more famously, Archimedes used this method to arrive at his approximation of pi—the constant that is a factor in both the area and circumference of circles. Essentially, the method of exhaustion is a process in which the geometer constructs a figure with a known area inside a figure with an unknown area (such as a square inside a circle). Then, the geometer constructs a second figure with a known area outside of the unknown figure (a second square, say, outside the same circle). The closer the interior and exterior figures adhere to the perimeter of the unknown figure, the closer the approximation of the central figure's area, which naturally lies somewhere between the two known areas.

Overall, in the the *Elements,* Euclid proved 465 propositions and provided 131 definitions. And while he certainly drew upon older works (the difference in difficulty between one book and the next might be testimony to this), the design and construction of the book are his own triumph.

The work, as mentioned, was an enormous success, lasting for more than two millennia. And although Euclid himself never used the word *geometry*, the word was added to his work very early on in its history, appearing on Latin translations within a couple of centuries.

In the fourth century CE, some seven hundred years after the *Elements* was written, Theon of Alexandria translated it. Though not the first translation of Euclid's work, Theon's focus on students set his translation apart and made the *Elements* much more accessible. This version was widely distributed, supplanting translations that adhered more closely to Euclid's prose. Until the nineteenth century, variations of Theon's translation were the only versions available.

Newton

The *Elements* was translated again and again from the ninth through thirteenth centuries in an effort to improve it. It was particularly popular in the Arab world; Euclid topped the reading lists of Arab mathematicians. Toward the end of the Middle Ages, Euclid's *Elements* was rediscovered by Western culture and was translated into English. The first English translation was titled *The Elements of Geometrie of the Most Ancient Philosopher Euclide of Megara.* At this point, the work was clearly better known that its author. In fact, the book was attributed to the wrong Euclid. The philosopher Euclid of Megara was a contemporary of Plato, living a hundred years before Euclid of Alexandria.

Nonetheless, the *Elements* is the most influential math text ever. More than a thousand different editions have been produced since the invention of the printing press, and only the Bible exceeds Euclid's masterwork in number of copies printed.

But the influence of Euclid's *Elements* extends far beyond mathematics and geometry. The ironclad nature of the text has inspired other writers to structure their work along the same lines. For example, Isaac Newton organized his *Principia Mathematica*, published in 1687, in the manner of

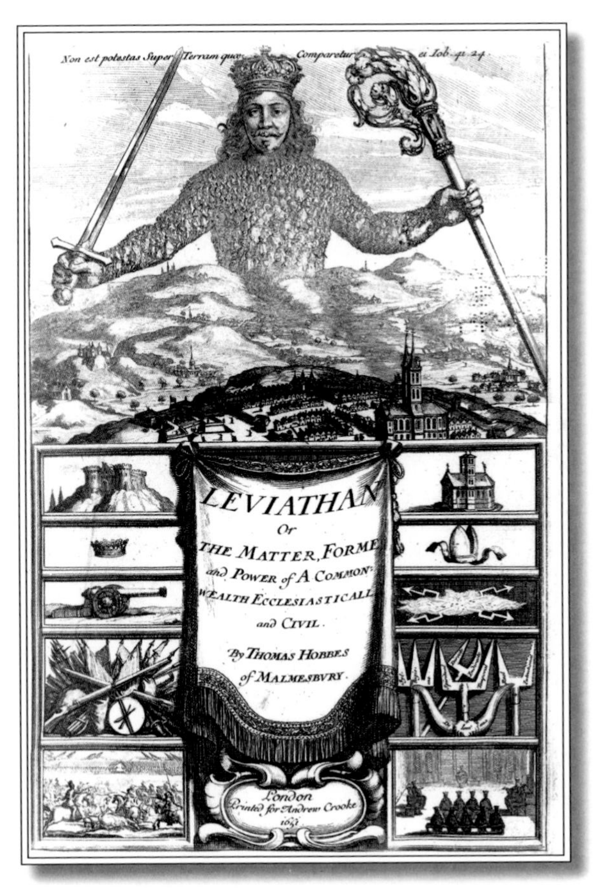

Philosopher Thomas Hobbes organized his *Leviathan* like Euclid's *Elements* in an effort to make his philosophical writings as undeniable as Euclid's mathematical proofs.

the *Elements*. Benedict de Spinoza's *Ethics* (1677) used the same format in an effort to deductively prove the existence of God. And, in writing *Leviathan* (1651), Thomas Hobbes expressly sought to mimic the organization of the *Elements* to make his political and moral philosophy as indisputable as Euclid's proofs. It is not just mathematicians who owe a great deal to Euclid. Thinkers in all disciplines are in his debt.

Uncertain Definitions

One feature of Euclid's *Elements* puzzles historians. At the beginning of many of the books are definitions of terms, and these definitions contain certain inconsistencies. Even more curiously, the definitions are never referred to within the main text. This has led some observers to wonder: Could the definitions have been written after the *Elements* and added to later editions during copying?

Historians J. J. O'Connor and E. F. Robertson explore this possibility in their paper "Euclid's Definitions." They note, for example, that Euclid defines certain common terms (such as *point* and *line*) twice, while entirely omitting definitions for other terms (such as *magnitude*). They suggest that Euclid's definitions might in fact be the work of Heron, a later Alexandrian geometer who is credited with a book titled *Definitions of Terms in Geometry.* Its definitions and those found in the *Elements* are very similar. Previously, the consensus was that Heron had lived in the third century CE, and many historians believed that he had adapted Euclid's definitions for his work. Because Euclid's definitions were referred to by Sextus Empiricus, who wrote in the late second century CE, attributing them to Heron was out of the question.

However, recently uncovered evidence has revealed that Heron actually lived in the first century CE, which means his *Definitions* came before Sextus wrote. Thus it is possible that Heron's definitions were appended to Euclid's *Elements* during the copying of scrolls by an expert aware of both works. Of course, it is also still possible that Heron adapted his definitions from Euclid's work.

Yet another possibility exists: that the definitions originate with a third, unknown source, and were adapted by both Heron and Euclid, or by one of the people copying the *Elements* in its long journey through time. The truth may never be known with certainty.

Timeline

ca. 325 BCE Euclid is born.

332 BCE Alexandria is founded by the Macedonian conqueror Alexander the Great.

323 BCE Alexander the Great dies.

306 BCE Ptolemy I takes control of Alexandria.

280 BCE Museum and Library of Alexandria founded.

ca. 265 BCE Euclid dies.

four
Archimedes

In the seventeenth century, Isaac Newton and Gottfried Wilhelm Leibniz, building on the work of many predecessors, independently developed calculus. Astoundingly, a Greek mathematician had worked out some of the fundamental principles of integral calculus nineteen hundred years earlier—and he had done so single-handedly. That mathematician, Archimedes, also made great breakthroughs in the areas of statics (how forces operate using simple machines, such as a lever), hydrostatics (how forces operate in water), geometry, and engineering. So incredible were his gifts for mathematics and physics that Archimedes' contemporaries dubbed him "the Master."

More than two millennia after his death, many of Archimedes' discoveries seem to be the product of common sense, but they instead resulted from his extraordinary insight. "The fact is," wrote the famous Greek philosopher and biographer Plutarch (ca. 46-ca. 119 CE), "that no amount of mental effort of his own would enable a man to hit upon the proof of one of Archimedes' theorems, and yet as soon as it is explained to him he feels that he might have discovered it himself."

During his time, Archimedes was renowned throughout the Mediterranean region. But his achievements would have been largely unknown to the modern world if not for the efforts of a German mathematician and astronomer called Johannes Regiomontanus. During the Renaissance,

Johannes Regiomontanus

Regiomontanus—whose real name was Johann Müller—discovered Archimedes' writings. He set to work translating the texts, planning to publish both the Greek originals and the translations. Although Regiomontanus died before his plan could reach fruition, his assistant, a printer, carried on for him and made the works available to the public

Nevertheless, many important details about Archimedes' life and work remain unknown—and many frequently told stories about him may or may not be true. It is not known, for instance, whether Archimedes ever married or had children. Nor can it be said with certainty whether the famous account of Archimedes running naked through the streets of his city after an important discovery is anything more than a fanciful tale. The picture of Archimedes that has come down through the ages is of an absentminded genius so engrossed in the world of ideas that he often forgot about his own physical needs. Plutarch relates that Archimedes often needed to be reminded to eat and sleep, and that sometimes his slaves had to pick him up and carry him to the bathhouse so he would bathe. Even then, however, he would remain absorbed in his work, drawing figures and diagrams in the soap bubbles, ashes, and oils used for cleansing. While this picture is vivid and certainly resonates with modern readers—the absentminded professor is a familiar figure today—it must be remembered that Plutarch wrote some two hundred years after Archimedes' death. It is impossible to know whether he had reliable sources or whether he made up the characterization.

Historians do know that Archimedes was born around the year 287 BCE in the Greek city-state of Syracuse, located on the island of Sicily. He died seventy-five years later, during a Roman invasion of Syracuse.

Eratosthenes of Cyrene

The son of an astronomer named Pheidias, Archimedes began learning astronomy at an early age. He traveled to Alexandria, where he was educated at the city's famous Museum. There he may have been instructed by Euclid, though it is more likely that his teachers were successors to the renowned author of the *Elements*. In any event, Archimedes did take a Euclidean approach to his work, setting down general principles and then deducing every step afterward.

Archimedes is known to have had four notable friends. The first, King Hieron II, was the king of Syracuse. He may have played a role in Archimedes' initial rise to prominence, for even a genius of the Master's caliber could use a leg up. Archimedes also counted as close friends three fellow scientists who taught at the Museum at Alexandria: the astronomer Conon of Samos and two geometers, Dositheus and Eratosthenes of Cyrene. After finishing a book, Archimedes would forward it, along with a letter, to these three men. This was tantamount to publishing the books, as the manuscripts would be turned over to the Library of Alexandria and copied for the collection. Archimedes published numerous books, including *On Spirals, On the Measurement of the Circle,*

Quadrature of the Parabola, On Floating Bodies, Sand Reckoner, and *On the Sphere and the Cylinder.*

Sometimes Archimedes also sent his friends a mathematical proposition that he had proven—but without including the proof, in order to give them the pleasure of puzzling it out for themselves. Some of these theorems made their way around the mathematical community. On a few occasions, an unscrupulous mathematician would announce one of Archimedes' theorems as his own discovery. To trip these liars up, Archimedes began sending an occasional false proposition, which he would quickly demonstrate to be erroneous if someone else claimed to have discovered it.

An illustration of Archimedes' method of determining the area of a parabolic section by dividing it into smaller and smaller triangles.

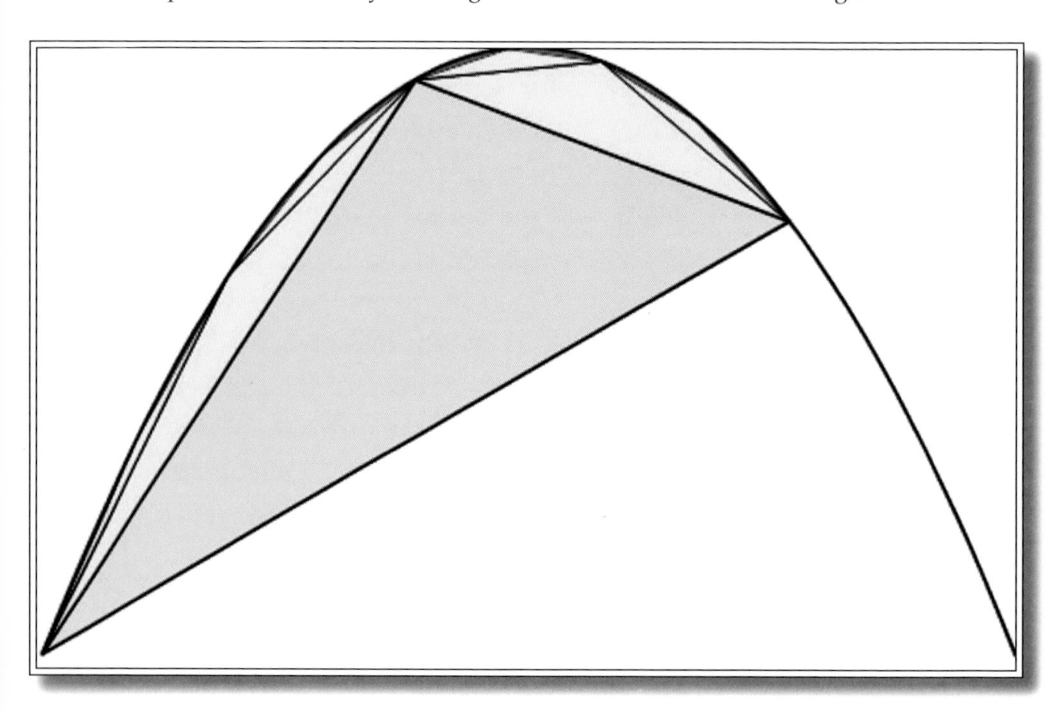

One of Archimedes' most impressive achievements in geometry was detailed in his book *Quadrature of the Parabola.* In it, he showed the way to determine the area of a parabolic section (a curve in the shape of a parabola, but with endpoints somewhere along the curve). First, he drew a line segment, or chord, between the two endpoints of the parabola. Then, parallel to the chord, he drew a second line, passing through a single point of the parabola. (This line is called a tangent.) Once the tangent point was found, Archimedes constructed a triangle with vertices at the two endpoints and the tangent point. Archimedes gave this triangle a value of 1.

This triangle did not take up the entire space of the parabolic section, however. It left unaccounted for two curves—smaller parabolic sections—between the tangent and the two endpoints. With each of these sections, Archimedes repeated the same procedure, finding the tangent points and constructing two smaller triangles. This second step left four even smaller parabolic sections unaccounted for, but Archimedes could now find the area of the two new triangles. Through geometry, he was able to determine that the sum of the area of these triangles was one-quarter of the area of the original triangle.

Archimedes was able to repeat this with the four parabolic sections left over from this second pass, and he obtained the same result: each pair of new triangles was equal to one-quarter the area of its "parent" triangle. This relationship held true for these four, and the eight triangles they led to, and the sixteen triangles after that, and the following thirty-two, and so on. And as he proceeded in this manner, Archimedes was also able to show that the error—the leftover parabolic sections—was predictably getting smaller with each step.

Since the error was approaching zero and the areas of the new triangles were predictable, Archimedes was able to calculate the area of the section as $1 + 1/4 + 1/4^2 + 1/4^3 + 1/4^4 \ldots + 1/4^n$, where n = the number of stages of triangle construction. Taken to its ultimate end, this meant that the area of parabolic sections was 4/3 the area of the initial triangle constructed from its endpoints and its parallel tangent. It was Archimedes' groundbreaking notion of measuring smaller and smaller segments that provided the intellectual foundation for integral calculus. His execution of this concept is why many people consider him the father of the discipline.

Many people also consider Archimedes the father of physics. Unlike his predecessors, including Pythagoras and Euclid, Archimedes applied scientific principles to real-world problems. He also conducted physical experiments, though he did not regard a successful experiment as proof of a given theorem. Instead, he used experimentation—what he called the "mechanical method"—to point him in the proper direction in his search for mathematical or geometrical proof.

One of Archimedes' methods of proving things was called *reductio ad absurdum,* Latin for "reduction to absurdity." This method consisted of extrapolating a set of conditions to an extreme point such that it becomes impossible to accept. But in order to do so, Archimedes needed to know in which direction he should extrapolate the data. Otherwise he could continue working outward trying to disprove something genuinely true. Experimentation showed him the correct direction to go.

Perhaps the most famous story about Archimedes involves an incident in which an ordinary activity led him not only to the solution to a vexing problem but also to the discovery of a basic physical principle. The story, first told by Vitruvius, writing in the first century BCE, goes that Archimedes' friend King Hieron II had been presented with a ceremonial wreath. The wreath was supposedly made of solid gold, but Hieron suspected it was actually gold alloyed with silver or some other cheaper metal. Hieron asked Archimedes how he could tell if his goldsmith had cheated him.

While pondering how to differentiate between the two metals, Archimedes went to take a bath. As he lowered himself into the bath, he noticed the water level rising to accommodate him. This sparked an idea, and he supposedly reacted with such enthusiasm that he left the bathhouse and ran naked down the street, shouting "Eureka! Eureka!"—meaning "I've found it! I've found it!"

The incident sounds almost too good to be true, tying Archimedes' genius, his absentmindedness, and even his reluctance to bathe in a neat little mythological bow. Yet whatever the truth of the story, the fact remains that Archimedes originated a principle explaining how forces act on floating bodies. In simple terms this principle, called the law of buoyancy (or, alternatively, Archimedes' principle), states that an object will float if it has a lower specific gravity (or density) than the fluid it displaces by its volume. A rock, for example, will sink because the water it displaces weighs less than the rock itself. A log will float, on the other hand, because the water it displaces weighs more than the log. Ships can float, even when made of dense metals, because they are hollow inside: the wide hull displaces vastly more water than the metal would if crushed into a solid block. Therefore, the ship's average density is less than that of the water.

A sixteenth-century illustration of Archimedes discovering the law of buoyancy. *(Courtesy of The Print Collector/Alamy)*

Archimedes realized that if he could weigh Hieron's wreath as it was submerged, he could discover the density of the metal used to make it. Apparently, he was able to construct a hydrostatic balance (a design of one such instrument found in the second century CE is attributed to Archimedes) and test the wreath. Discovering the wreath's density to be less than an object he knew to be pure gold, Archimedes was able to prove that the wreath was made with inferior materials, and that the goldsmith was a fraud. In his book *On Floating Bodies,* Archimedes presented the law of buoyancy, as well as other principles of hydrostatics.

Another of Archimedes' greatest achievements was discovering an excellent approximation of π (pi), a constant needed to determine the area and circumference of a circle (as well as a sphere and cone). To do this, Archimedes used Eudoxus's principle of exhaustion. That is, he inscribed a polygon within the perimeter of the circle, and circumscribed a second polygon on the outside of the perimeter. If he could calculate the area of each of those polygons, Archimedes would know that the perimeter of the circle lay somewhere between those two perimeters.

Archimedes began with two hexagons, one within the circle and the other surrounding it. He would calculate the perimeters, then double the sides of the polygons, going from the six-sided figures up to shapes with twelve, twenty-four, forty-eight, and finally ninety-six sides. As the polygons gained more sides, they adhered more closely to the shape of the circle they fenced in, and the values of their perimeters crept closer and closer to the circumference of the circle itself. As the difference between the three figures became smaller, the room for error also diminished, making the approximation more accurate.

Archimedes then took the perimeters of both ninety-six-sided polygons and set them as the upper and lower boundaries for the circumference of the circle. A circle's circumference is the product of π and its diameter, so by dividing the perimeter of the interior polygon by the circle's diameter, Archimedes was able to obtain a lower-bound value for π of

3 10/71. By repeating the process with the exterior polygon, he got an upper-bound value for π of 3 1/7. In decimal notation (something unknown in Archimedes' time), these values would translate to 3.14084 and 3.142858. Modern mathematicians have computed the value of π to 10 billion decimal places, the first 10 of which are 3.1415926535.

In 1896, a manuscript by Heron of Alexandria from approximately 60 CE was discovered. Titled the *Metrica,* it reported that Archimedes had come even closer to the true value of π in a later effort that had been lost. According to Heron, Archimedes apparently bounded π as greater than 211875/67441 and less than a variable called *u.* Heron later identified the variable as 197888/62351, but this makes no sense. With a decimal value of 3.1738, this is a larger number than Archimedes' earlier upper boundary. It is possible that this value for *u* is the result of a copying error, and the original figure is lost to time. Judging from the lower boundary, however, experts think the figure might have come from the perimeter of a polygon with an incredible 640 sides.

As amazing as his work with π truly was, Archimedes did not think it his greatest achievement. That he reserved for his discovery of the ratio between the volume of a sphere and that of a cylinder circumscribed around it. (This cylinder would share the sphere's radius for its circular sides, and use the sphere's diameter—twice its radius—as its height.) As detailed in his book *On the Sphere and the Cylinder,* Archimedes discovered that the ratio of a cylinder's volume to its inscribed sphere is 3:2.

Archimedes' many other remarkable achievements included his development of a mathematical notation to express extremely large numbers, which he set forth in

An Archimedes' screw *(Courtesy of Time Life Pictures/Mansell/Time Life Pictures/Getty Images)*

Sand Reckoner. This notation, he claimed, could express the number of grains of sand needed to fill the entire universe. Archimedes also invented an instrument to measure the angle of the sun, as well as a water clock to keep accurate time when the sun was down, rendering a sundial useless. Heron wrote that Archimedes discovered a rule for finding the area of a triangle from the lengths of its three sides. He developed a model planetarium and invented a device, called the Archimedes screw, to move water up an incline.

The Archimedes Screw

One of Archimedes' most useful inventions was an ingenious device used for lifting water up an incline. This invention, called the Archimedes screw, was—and still is—used to transport water for irrigation.

The construction of the machine is fairly simple. A large screwlike device is housed inside a hollow tube. The spiral blades of the screw fit fairly snugly inside the tube, although if the tube itself is stationary, the screw should still be able to move within it.

The bottom of the screw and tube should be placed in water, with the rest of the device leading uphill. Water should flow into the tube, and over the first few blades of the screw. Meanwhile, more water will flow into the bottom of the tube and cover the lower blades of the screw. After regular turning, the water will spiral around the screw, lifted higher and higher until it reaches its destination. The screw doesn't have to be completely watertight within the tube; it turns too fast to let much dribble down between the screw and the tube, and any water that seeps between the blade and the casing will be supported by the water beneath it.

Another method of construction allows the screw itself to attach to the inside of the tube casing, if the entire tube is free to move. There are drawings showing people turning large Archimedes screws by running on the tube itself, like a lumberjack rolling logs down a river. This kept the tube (and the attached screw) turning, and kept the water flowing uphill.

Archimedes made incredible breakthroughs in the field of statics, the study of the effect of forces on objects through the use of levers and other simple machines. His most influential achievement in this field was discovering an object's barycenter, also called its center of gravity (although Archimedes never used either term). The center of gravity of a given object is the point at which it will maintain its position relative to the horizon if it is supported by that point alone. In other words, it is the balance point, a place at which an object can be suspended or supported without tilting or wobbling. Archimedes determined the barycenters of various geometric figures, including parallelograms (the intersection of the lines bisecting the parallel sides) and triangles (the intersection of the median lines). He also used these results to determine the barycenters of solids.

This was interesting in and of itself, but Archimedes gave it a greater practical purpose in the use of the lever. A lever consists of a straight bar that rests on and pivots around a fulcrum, or support. It may be used to help lift heavy objects.

The longer a lever is, the more it multiplies the force applied to it. For example, if a five-pound weight rests on

An illustration of Archimedes moving the world with a simple lever.

a lever one foot away from the fulcrum, the force on the other side required to lift that weight would differ according to its distance from the fulcrum. At five feet away from the fulcrum, one pound of force would move the five-pound weight. At ten feet away, it would take only half a pound of force to move the weight. The greater the distance, the less force needed. This correlation is what Archimedes meant by "inversely proportional." Upon demonstrating this principle, Archimedes is said to have boasted, "Give me a place to stand, and I can move the Earth."

Archimedes is credited with the invention of the compound pulley system, though in the eighth century BC the Assyrians may have already possessed a pulley type of device. In any event, the compound pulley system made use of an idea similar to the leverage principle. The user of a compound pulley pulls a rope that wraps around and turns a wheel of a small circumference. This ring is attached to a larger

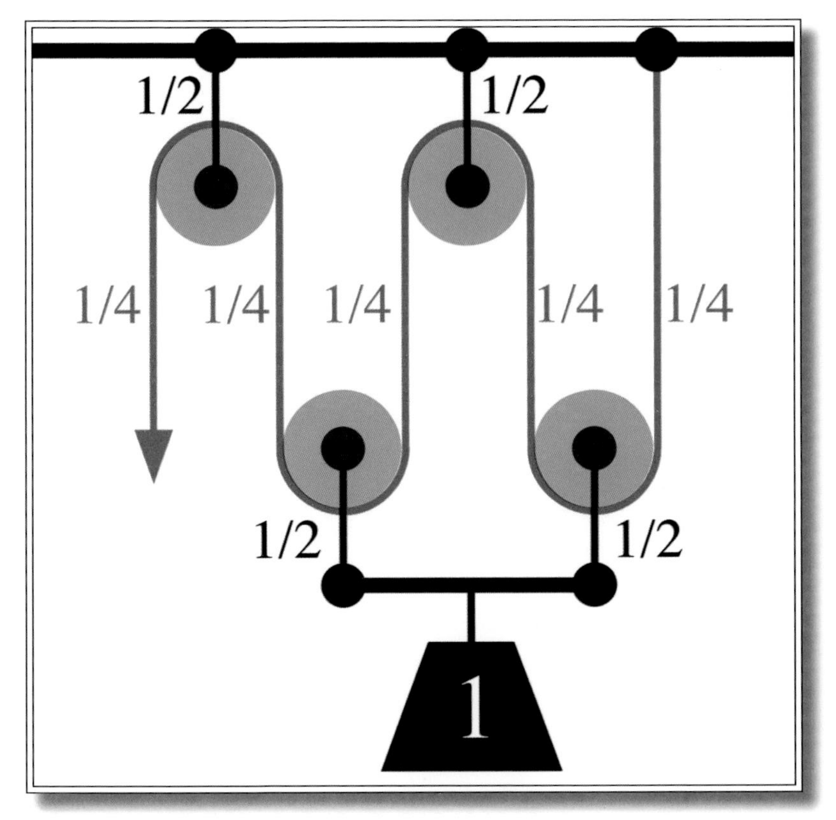

A compound pulley system

ring, which turns at the same rate. This has a greater effect of motion, since its circumference is larger, which causes a second rope, wrapped around it, to move with greater speed. Archimedes demonstrated the strength of compound pulleys to King Hieron by pulling a boat into harbor.

King Hieron had a use for this technology. He commissioned Archimedes to design weapons of war for the defense of Syracuse. It was a prescient decision: After Hieron's death in 215 BCE, these machines proved invaluable. Two major powers, Rome and Carthage, were engaged in a struggle called the Second Punic War, and Syracuse had allied itself with the

Carthaginians. In response the Romans put the island under siege. The city-state was facing down a much larger force than it could muster. On the Romans' side was a powerful navy. On Syracuse's side was Archimedes.

Archimedes used his knowledge of levers and pulleys to construct fearsome siege engines to protect the city. These included catapults to hurl missiles at the Roman ships. Before Archimedes, catapults were of limited use against moving targets, because they were built to fire at one range and one range only. Archimedes designed catapults with adjustable ranges, which would have better success against a mobile opponent. The catapults were capable of hurling rocks as heavy as five hundred pounds.

In this engraving, Archimedes' machines are being used to defend the island of Syracuse against Roman attack. *(Courtesy of HIP/Art Resource)*

Syracuse was situated right on the coast of Sicily. Although it was protected by cliffs and high walls, enemy ships could sail almost to the city's edge. Archimedes prevented their approach with levers and pulleys. He had constructed giant levers to swing ropes with grappling hooks down onto the Roman ships. These claws would take hold, and then the operators of the machines could use leverage to pull the ships out of the water. Plutarch described the terror these devices instilled: "At last the Romans were reduced to such a state of alarm that if they saw so much as a length of rope or a piece of timber appear over the top of the wall, it was enough to make them cry out, 'Look, Archimedes is aiming one of his machines at us!' and they would turn their backs and run."

Archimedes supposedly used one more method to defend the city. Although accounts of this invention are considerably less plausible to modern scholars, the legend persists. Archimedes is said to have built what amounted to a solar-powered death ray to light Roman ships on fire. Different sources describe the device in different ways. One source describes a hinged mirror Archimedes built to reflect and focus the sun's light. Another story says that Archimedes instructed warriors to polish their shields and arrange them so that they all reflected the sun at the same point on an enemy ship. But however Archimedes focused the sunlight, the end result was the same: the concentrated solar heat set Roman ships ablaze.

This story raises a number of red flags. For one thing, the first description of this amazing weapon is dated around 160 CE, almost four hundred years after the battle. Also, if a long-range armament with such deadly effectiveness had been invented, why had no one used it since? Both of these points suggest that the death ray was a fiction concocted by someone embellishing Archimedes' reputation.

Nonetheless, modern scientists have tested the principle, with mixed results. In 1973, Dr. Ioannis Sakkas enlisted the Greek navy to help him with an experiment. Using seventy sailors holding large shields to reflect the light, Sakkas was apparently able to set fire to a model ship about 160 feet away. But later experimenters, led by Adam Savage and Jamie Hyneman of the Discovery Channel's *Mythbusters* television show, were unable to reproduce his results on two different attempts.

With the help of Archimedes' machines, the defenders of Syracuse held off the Romans for three years. Eventually, however, the city's defenses were overcome. By this time, the story goes, Archimedes was already wrapped up in another project. It is said that the famously absentminded mathematician did not even notice the invasion was taking place. Supposedly, the Roman soldiers were ordered not to harm Archimedes but instead to bring him to their commander. Archimedes was engrossed in a geometrical problem, according to one story, when a Roman soldier arrived at his house. "Don't disturb my circles," the mathematician implored. The soldier responded by striking him down.

Archimedes had asked that the accomplishment of which he was proudest—his discovery of the 3:2 ratio between the volume of a sphere and the volume of a cylinder circumscribed around it—be commemorated on his tombstone. Apparently his wishes were carried out. While serving as an administrator in Sicily around 75 BCE, the Roman orator and statesman Cicero searched for Archimedes' tomb. After some effort, he located a grave "completely surrounded and hidden by bushes of brambles and thorns." Only a column "surmounted by a sphere and a cylinder" rose above the vegetation. After the weeds had been cleared away, a marker inscribed with some still-legible verse confirmed to Cicero that he had indeed found the final resting place of Archimedes.

Just as Cicero had to clear away weeds and brambles from Archimedes' tomb, some of Archimedes' writing was obscured through the centuries. One work, *The Method,* was lost until 1899, when it was discovered as a palimpsest. A palimpsest is a document containing one text recorded over the erased remains of another. When paper was rare or expensive to produce—as, for example, during the Middle Ages—older texts considered to have little value were routinely obliterated so that the paper could be reused. The pages of several works by Archimedes, including *The Method* and the only surviving copy of *On Floating Bodies* in Greek, were bleached by medieval monks, bound together, and written over to form a prayer book in 1229. By 1906, it was confirmed that the original work beneath the prayer book was that of Archimedes. John Ludwig Heiberg, the world's leading Archimedes scholar, studied the work closely. Although the Archimedes palimpsest disappeared in 1908, Heiberg published his findings between 1910 and 1915.

The document was long thought to be lost forever. Then, unexpectedly, it turned up for sale at the famous auction house Christie's in 1998. An anonymous American collector bought the book, then lent it to the Walters Art Museum in Baltimore. The institution has been studying the work since 1999 in an effort to faithfully reproduce Archimedes' original text.

Timeline

87 BCE Archimedes is born in Syracuse, on the island of Sicily.

215 BCE The Second Punic War begins, and when Syracuse supports Carthage, the Romans decide to invade.

212 BCE Archimedes is killed by a Roman soldier.

five
Hypatia

Nearly six hundred years after the death of Archimedes, around 370 CE, a daughter was born to a mathematician named Theon of Alexandria. Theon vowed to raise the girl as the perfect human being. This was unusual, not only for its ambition, but because at the time women were not considered the equals of men. Nonetheless, Theon gave his daughter Hypatia the best education he could. And he could provide for quite a lot, since he was a teacher at the famed Museum at Alexandria.

Alexandria: Center of Learning

The city of Alexandria was founded around 332 BCE by the Macedonian conqueror Alexander the Great. Situated on the coast of the Mediterranean Sea at the mouth of the Nile River, the city was not originally part of Egypt proper. Instead, it was treated as its own state, much like the Vatican within Rome. Within ten years of the city's founding, Alexander died, leaving his generals to squabble over his vast empire. By 306, some of the dust began to clear and Alexandria emerged under the control of Ptolemy I, the first of a long dynasty of rulers for the city in its golden age.

The Ptolemy line—particularly the first three rulers—established the city as a center of learning and culture. The first Ptolemy began building an enormous museum in the city. At the time, museums were primarily educational institutions, more like an academy or university than a building for displaying art or historical artifacts. The Museum of Alexandria drew a diverse academic community of Greek, Egyptian, and Hebrew scholars.

The Museum employed many famous thinkers as teachers, including Euclid and Hypatia. Its chief librarian for a time, Eratosthenes (273-192 BCE), was also acclaimed for his mathematical and astronomical work. A contemporary of Archimedes, he was the first person to successfully calculate the circumference of the earth. He did this by comparing the sun at its zenith in two places a known distance apart. From this information, he achieved a result only 5 percent different from the figure accepted today. Eratosthenes is also said to have calculated the distance from earth to the moon and the sun. The distance to the sun he arrived at holds up well against modern science, but he was wrong about the distance to the moon.

The Museum of Alexandria would also house a great library, a project that began to take shape under the reign of Ptolemy II. The library was filled in several ways. Ptolemy bought as many books as he could. (These books were actually rolls of papyrus about ten to fifteen feet long, which held anywhere from 10,000 to 20,000 words on them.) Also, travelers into Alexandria were forced to give up any books they brought with them at the city's edge. The books would be taken to the library and copied. Then the copies would be returned to the owner, while the originals stayed in the Museum. Sometimes Ptolemy II took matters even further, asking to borrow books from people in neighboring lands, and leaving a deposit for each book's return. Then he would take the book, copying it if necessary, and forfeit the deposit. No matter what the aggrieved owners might have thought, the fees were a small price for him to pay for building the largest library the world had ever known.

Ancient scholars using the library of Alexandria *(Courtesy of North Wind Picture Archives/Alamy)*

The Museum—at the time, the word was almost synonymous with "research institute"—was built around 280 BCE and maintained throughout the centuries by the Ptolemy dynasty that ruled Egypt, beginning with Ptolemy I Soter. Although many of the buildings of the Museum were destroyed in a civil war in 272 CE, the institution continued on with its mission. Theon, a teacher of mathematics and astronomy, was one of the last members of its faculty.

Theon educated Hypatia in both of his specialties at the Museum. He also brought her with him on his travels, exposing her to great teachers in Athens and Italy. In addition to her aptitude for mathematics and science, Hypatia also displayed a gift for philosophy.

In time, Hypatia became a renowned intellectual in her own right. Socrates Scholasticus, a fifth-century scholar who should not be confused with his more famous namesake, wrote in his *Ecclesiastical History* that Hypatia "made such attainments in literature and science, as to far surpass all the philosophers of her own time." Upon reaching adulthood, Hypatia gained a position teaching math and philosophy at the Museum in Alexandria. The city also appointed her to be a municipal chair of philosophy.

Hypatia's writing reflected her impressive learning. For example, her commentary on the book *Arithmetica,* an algebra text written by the third-century CE mathematician Diophantus, was so insightful—and so well received by the intellectual community—that much of it was incorporated into all later editions of the original work.

She also collaborated with her father. Together, Theon and Hypatia worked on a new edition of Euclid's *Elements,* which they hoped would be more accessible to beginning students. They also coauthored a separate treatise on Euclid, and they worked together on at least one book on the second century CE astronomer and mathematician Ptolemy. Theon's inscription at the beginning of his commentary on Ptolemy's *Almagest*—a book detailing the motions of the stars and planets—suggests that Hypatia wrote the third chapter of his commentary on that work as well. She was an avid astronomer in her own right.

Like Euclid before her, Hypatia was fascinated by the properties of conic sections, the curved shapes that result from cutting a cross section of a cone. Hypatia devoted much of her time and energy to their study. She wrote *On the Conics of Apollonius,* a commentary on the work of Apollonius of Perga, who lived around Archimedes' time. Apollonius had examined the irregular orbits of planets. Hypatia's commentary popularized his work six hundred years after his death. Hers was the last major work on conic sections until the eleventh century.

Hypatia's interest in astronomy led her to come up with her own design for an ancient instrument known as the plane astrolabe. This device allows the user to determine the positions of the sun and stars in the sky. It consists of a disc, called a mater, with a hollow space in the center that can fit a round plate, called a tympan. The tympan, which is engraved with the positions of the stars, can be rotated to match what the night sky looks like from the user's perspective. Thereafter, it should be rotated every hour as the stars move. With that information, users can determine their latitude and the time of day. The device is particularly useful to sailors at sea, where there are no landmarks to aid in navigation.

The fact that Hypatia designed an astrolabe is known

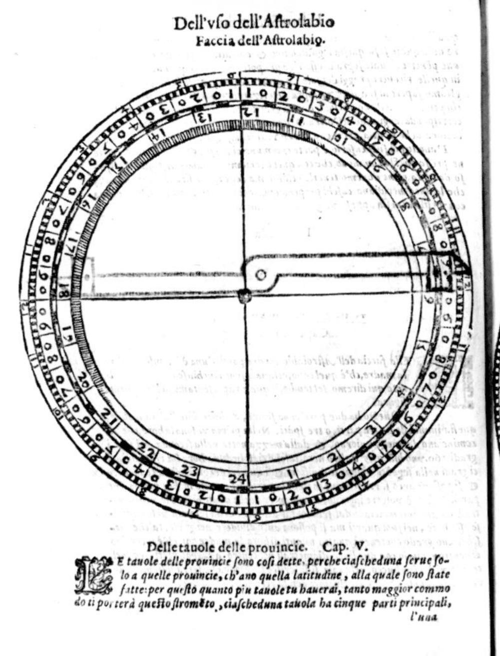

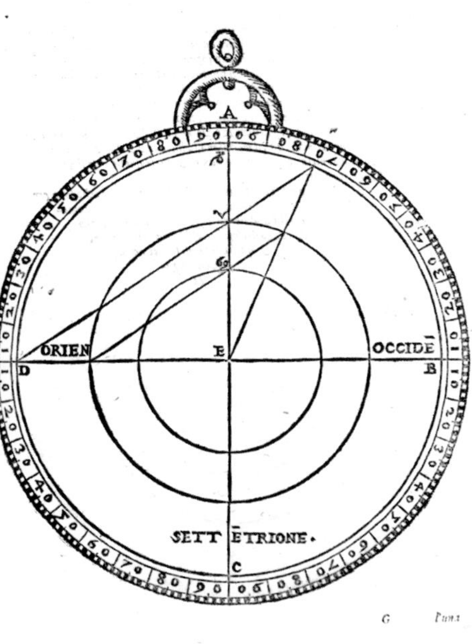

Two views of a sixteenth-century astrolabe *(Library of Congress)*

from letters written to her by one of her students, Synesius of Cyrene. Some historians believe Synesius was a pagan who later converted to Christianity; others think he was always a Christian. In any case, he eventually became bishop of Ptolemais. Yet Synesius was also a Neoplatonist—an adherent to a philosophical system that adopted ideas not only from Plato, but also from later philosophers such as Aristotle and from Eastern religious traditions. Some Christians considered Neoplatonism opposed to their religion.

Six of Synesius's letters to Hypatia have survived, and several other letters he wrote mention her. From his letters, many scholars have concluded that Hypatia invented a variety of scientific instruments, including a device to measure water level, a device to distill water, and a device to measure the density of liquid, which she called a "hydroscope" and which the seventeenth-century mathematician Pierre de Fermat later identified as a hydrometer. Synesius had asked for this device specifically, claiming he needed it to help him through what turned out to be a fatal illness. He probably needed the hydrometer to measure the proper dosage of medicine, although it is possible that he intended to use it to test his own fluids.

Synesius was not the only student devoted to Hypatia. She was a very popular teacher in Alexandria. Often, she conducted teaching sessions in her own home, where students would gather to discuss and debate scientific and philosophical questions. During Hypatia's time, Christianity had begun gaining ground in Alexandria, and pagans were not permitted to attend schools set up for Christians. But all students—whether Christian, Jew, or pagan—were welcome at Hypatia's lessons.

In his *Life of Isidorus,* the Neoplatonist philosopher Damascius recounts that Hypatia would sometimes "put on her philosopher's cloak" and walk through town as people questioned her about Plato, Aristotle, or any other philosopher. In this way, her reputation spread, even among people who were not studying at the Museum.

Hypatia's philosophical beliefs were influenced by the third century CE Roman philosopher Plotinus and by Iamblichus (who wrote *On the Pythagorean Life* in the early fourth century). Both, like her friend Synesius, were Neoplatonists. Hypatia herself may or may not have been a Neoplatonist, but many people in Alexandria assumed she was.

Hypatia enjoyed a high public standing, and she was very active in Alexandrian politics. City magistrates often consulted her, and ambassadors from other countries paid her visits.

Hypatia was also well known for her virtue and chastity. She never married, and she is said never to have taken a lover.

As her fame grew, friction between Hypatia and her city began to increase. She was a pagan, and Alexandria was becoming increasingly Christian. In 412, a fanatical Christian named Cyril became bishop of Alexandria. Cyril persecuted the city's Jews, killing thousands. He also

Cyril *(Courtesy of Hulton Archive/Getty Images)*

wanted to drive Neoplatonists and pagans away. His intolerance naturally caused a rift between Hypatia and the city government.

On top of that, Cyril entered into a feud with Orestes, the Roman prefect of Egypt. Orestes also was one of Hypatia's students, and a friend as well.

Cyril's anti-Neoplatonist speeches became increasingly fiery over time, and he stirred up much hatred in the city. It remains unclear if he had a direct hand in what happened next, but most accounts say his speeches set the stage for the brutality that followed.

One March night in the year 415, Hypatia was on her way home when her carriage was stopped by a mob. She was pulled from the vehicle and dragged into a church called the Caesarium. Under the leadership of a magistrate named Peter, the mob stripped her naked and held her down. Then they scraped away her flesh with brick tiles. (Some historians say they used oyster shells, because the Greek word Socrates Scholasticus uses is *ostrakois*—literally "oyster shells"—but this was also the term for brick roofing tiles used in the city.) After Hypatia was dead, they cut her body into pieces and brought the pieces into a square, called Cinaron, where they burned the remains for the town to see.

Orestes made an official report of the murder and requested an investigation. Despite his urgings, the investigation kept being postponed for a supposed lack of witnesses. Eventually, Cyril claimed that Hypatia was alive and had moved to Athens. Even if he had no connection to the mob, he clearly was unwilling to punish those responsible for removing one of his political enemies.

Hypatia's death, in the view of some scholars, marks the end of an extremely fertile era in the history of mathematics. In the more than nine hundred years separating her lifetime from that of Pythagoras, the ancient Greeks made remarkable advancements, especially in the area of geometry. After Hypatia, the Greek mathematical tradition stagnated.

But Hypatia's life also marked a beginning. She is the first prominent female mathematician and scientist known to history, and her story continues to inspire people today.

Timeline

370 CE Born in Alexandria.

ca 400 CE Writes commentaries on Diophantus and Apollonius.

415 CE Murdered by a Christian mob.

Sources

CHAPTER ONE: Mathematics in Ancient Greek

p. 20, "Let no one ignorant . . ." Fred L. Wilson, "Science and Human Values: Plato," http://www.rit.edu/~flwstv/plato.html.

CHAPTER TWO: Pythagoras

p. 29, "Albert Einstein of his day," Estelle A. DeLacy, *Euclid and Geometry* (New York: Franklin Watts, 1963), 29.

p. 34, "The Samians were not . . ." J. J. O'Connor and E. F. Robertson, "Pythagoras of Samos," MacTutor History of Mathematics Archive, http://www-history.mcs.st-and.ac.uk/~history/Biographies/Pythagoras.html.

p. 34, "Stop, don't keep beating . . ." Jacques Brunschwig and Geoffrey E. R. Lloyd, eds., "Pythagoreanism," *Greek Thought: A Guide to Classsical Knowledge*, trans. Catherine Porter (Cambridge, MA: Belknap Press of Harvard University Press, 2000), 920.

p. 36, "Pythagoras especially valued . . ." Ibid., 923.

p. 36, "especially made use of it," Ibid.

p. 39, "whenever [Pythagoras] reached out . . ." Ibid., 921.

p. 39, "the chief of swindlers," Ibid., 920.

p. 40, "When Pythagoras found . . ." Carl Huffman, "Pythagoras," *Stanford Encyclopedia of Philosophy,* http://plato.stanford.edu/entries/pythagoras/.

p. 42, "A figure and a platform . . ." O'Connor and Robertson, "Pythagoras of Samos."

CHAPTER THREE: Euclid

p. 58, "to draw a straight line . . ." Web site of David E. Joyce, professor of mathematics and computer science, Clarke University, "Euclid's Elements: Book I," http://aleph0.clarku.edu/~djoyce/java/elements/bookI/bookI.html.

p. 58, "to produce a finite . . ." Ibid.

p. 58, "to describe a circle . . ." Ibid.

p. 58, "All right angles equal . . ." Ibid.

p. 59, "When a straight line standing . . ." Ibid.

p. 59, "If a straight line falling . . ." Ibid.

p. 62, "Things which equal the same . . ." Ibid.

p. 62, "If equals are added . . ." Ibid.

p. 62, "If equals are subtracted . . ." Ibid.

p. 62, "Things which coincide . . ." Ibid.

p. 62, "The whole is greater . . ." Ibid.

p. 62, "To construct an equilateral triangle . . ." Ibid.

p. 64, "Euclid is not the father . . ." Petr Beckmann, *A History of Pi* (New York: Barnes & Noble Books, 1993), 48.

p. 65, "there was no royal road . . ." O'Connor and E. F. Robertson, "Euclid of Alexandria," MacTutor History of Mathematics Archive, http://www-history.mcs.st-and.ac.uk/~history/Biographies/Euclid.html.

p. 67, "that by virtue of which . . ." Web site of David E. Joyce, professor of mathematics and computer science, Clarke University, "Euclid's Elements: Book VII, http://aleph0.clarku.edu/~djoyce/java/elements/bookVII/book VII.html" \l "defs.

p. 67, "a multitude composed . . ." Ibid.

p. 67, "that which is divisible . . ." Ibid.

p. 67, "that which is not divisible . . ." Ibid.

p. 67, "that which is measured . . ." Ibid.

p. 67, "the greater number . . ." Ibid.

CHAPTER FOUR: Archimedes

p. 74-75, "The fact is . . ." Melvyn Bragg, *On Giants' Shoulders* (New York: John Wiley & Sons, 1999), 19.

p. 88, "Give me a place to stand . . ." Sherman Stein, *Archimedes: What Did He Do Besides Cry Eureka?* (Washington, DC: Mathematical Association of America, 1999), 5.

p. 90 , "At last the Romans . . ." Stuart Hollingdale, *Makers of Mathematics* (London: Penguin, 1994), 66.

p. 92, "Don't disturb my circles," Stein, *Archimedes*, 3.

p. 93, "completely surrounded and hidden . . ." Web site of Bradley W. Carroll, chair, department of physics at Weber State University, "From Cicero, *Tusculan Disputations*," http://physics.weber.edu/carroll/Archimedes/cicero.htm.

p. 93, "surmounted by a sphere . . ." Ibid.

CHAPTER FIVE: Hypatia

p. 99, "made such attainments . . ." Socrates Scholasticus, "Of Hypatia the Female Philosopher," (Ecclesiastical History), bk. VI, chap. 15, Internet Medieval Sourcebook, Fordham University Center for Medieval Studies, ed. Paul Halsall, http://www.fordham.edu/halsall/source/hypatia.html.

p. 102, "put on her philosopher's cloak," Damascius, "The Life of Hypatia," *The Life of Isidore*, (reproduced in *The Suda*), trans. Jeremiah Reedy, http://cosmopolis.com/alexandria/hypatia-bio-suda.html.

Bibliography

Amos, H. D., and A. G. P. Lang. *These Were The Greeks.* Chester Springs, PA: Dufour Editions Inc., 1996.

Beckmann, Petr. *A History of Pi.* New York: Barnes & Noble Books, 1993.

Bragg, Melvyn. *On Giants' Shoulders.* New York: John Wiley & Sons, 1999.

Brunschwig, Jacques, and Geoffrey E. R. Lloyd, eds. *Greek Thought: A Guide to Classical Knowledge.* Cambridge, MA: Belknap Press, 2000.

DeLacy, Estelle A. *Euclid and Geometry.* New York: Franklin Watts, 1963.

Heath, T. L., ed. *The Works of Archimedes.* Mineola, NY: Dover Publications, 2002.

Hollingdale, Stuart. *Makers of Mathematics.* London: Penguin, 1994.

Ifrah, Georges. *The Universal History of Numbers.* New York: John Wiley & Sons, 1994.

Mankiewicz, Richard. *The Story of Mathematics.* Princeton, NJ: Princeton University Press, 2004.

Mlodinow, Leonard. *Euclid's Window.* New York: The Free Press, 2001.

Stein, Sherman. *Archimedes: What Did He Do Besides Cry Eureka?* Washington, DC: Mathematical Association of America, 1999.

Web sites

http://www-history.mcs.st-and.ac.uk/~history/index.htm
The MacTutor History of Mathematics archive, maintained by the School of Mathematics and Physics at the University of St. Andrews, Scotland, offers biographies of famous mathematicians.

http://cosmopolis.com
Alexandria on the Web.

http://aleph0.clarku.edu/~djoyce/java/elements/toc.html
This Web site, maintained by David E. Joyce, professor of mathematics and computer science at Clark University, Worcester, Massachusetts, provides a full translation of Euclid's *Elements*, along with helpful explanatory notes.

http://plato.stanford.edu/
The online *Stanford Encyclopedia of Philosophy* includes an entry on Pythagoras and Pythagoreanism, as well as numerous references to other Greek mathematicians.

Index